Ben Stacy Jerrik (Ed.)

Daniel W. Dobberpuhl

Ben Stacy Jerrik (Ed.)

Daniel W. Dobberpuhl

Electrical engineering, Microprocessor, United States Department of Defense

Part Press

Imprint

Publisher:
Part Press is a trademark of
International Book Market Service Ltd., 17 Rue Meldrum, Beau Bassin, 1713-01 Mauritius
Email: info@bookmarketservice.com
Website: www.bookmarketservice.com

Published in 2012

Printed in: U.S.A., U.K., Germany. This book was not produced in Mauritius.

ISBN: 978-613-6-27970-1

Contents

Articles

References

Daniel_W._Dobberpuhl

Daniel W. Dobberpuhl	
Born	March 25, 1945Streator, Illinois
Nationality	United States
Education	University of Illinois, Urbana-Champaign: bachelor of science in Electrical Engineering
Work	
Engineering discipline	Microprocessor design
Employer(s)	retired
Significant projects	MicroVAX, Alpha, StrongARM, PWRficient
Significant design	DEC Alpha
Significant awards	IEEE Solid-State Circuits Award (2003)

Daniel "Dan" W. Dobberpuhl (born 1945) is an electrical engineer in the United States who led several teams of microprocessor designers.

Background

Dobberpuhl was born in Streator, Illinois on March 25, 1945. He graduated with a bachelor of science degree in electrical engineering from the University of Illinois, Urbana-Champaign in 1967. He worked as an engineer for the Department of Defense until 1973 when he worked for GE Integrated Circuits Lab in Syracuse, New York, making application-specific integrated circuits.

In 1976 Dobberpuhl joined Digital Equipment Corporation (DEC) in Hudson, Massachusetts as a semiconductor engineer and led teams designing microprocessors such as the DEC T-11 and MicroVAX. He rose to become one of five senior corporate consulting engineers, DEC's highest technical positions.[1] As such he led the teams designing the first three generations of the DEC Alpha processor and published a 1985 text book called "a leading text in the field."[2]

He founded and directed the company's Palo Alto, California Design Center in 1993 where the StrongARM architecture was designed.

In 1998 Dobberpuhl co-founded SiByte, where as president he led the design of the SB1250 high performance MIPS System-on-a-chip processor. In 1998, EE Times named Dobberpuhl as one of the "40 forces to shape the future of the Semiconductor Industry." SiByte was funded by venture capital as well as large companies such as ATI Technologies, Cisco Systems, and Juniper Networks.[3] After a third round of $40 Million was raised in May 2000, the company was bought by Broadcom in November of that year for stock worth over $2 Billion.[4] Dobberpuhl stayed until 2003 as vice president and general manager of the Broadcom broadband processor division.

In 2003 the Institute of Electrical and Electronics Engineers named him a recipient of the IEEE Solid-State Circuits Award for "Pioneering design of high-speed and low-power microprocessors."[5] [6] Later that year he left to found P.A. Semi, a fabless semiconductor company that designed the PWRficient family of Power Architecture processors. He was elected to the National Academy of Engineering in 2006 for "Innovative design and implementation of high-performance, low-power microprocessors."[7] In 2008 P.A. Semi was sold to Apple Inc.

He received a Distinguished Alumni Award by University of Illinois in 2003,[8] and the College Of Engineering Alumni Honor Award from the University of Illinois in May 2009.[9]

Dobberpuhl retired from Apple near the end of 2009.[10] He holds 15 patents.[5]

Works

- Lance A. Glasser; Daniel W. Dobberpuhl (1985). *The design and analysis of VLSI circuits*. Addison Wesley. ISBN 978-0-201-12580-1.
- D.W. Dobberpuhl; et al. (November 1992). "A 200-MHz 64-b dual-issue CMOS microprocessor". *IEEE Journal of Solid-State Circuits* **27** (11): 1555–1567. doi:10.1109/4.165336.
- D.W. Dobberpuhl (September 15, 1997). "Circuits and technology for Digital's StrongARM and ALPHA microprocessors". *Seventeenth Conference on Advanced Research in VLSI*: 2–11. doi:10.1109/ARVLSI.1997.634842. ISBN 0-8186-7913-1.
- Committee on Sustaining Growth in Computing Performance (March 21, 2011). *The Future of Computing Performance: Game Over or Next Level?* [11]. National Academies Press. ISBN 978-0-309-15951-7.

References

[1] David Ditzel (December 5, 2003). "A Conversation with Dan Dobberpuhl; Topic: Power Management" (http://queue.acm.org/detail.cfm?id=957732). *ACM Queue*. . Retrieved July 2, 2011.

[2] "2003 - Daniel Dobberpuhl" (http://www.ieee.org/about/awards/bios/pederson_recipients.html#sect9). *IEEE Donald O. Pederson Award in Solid-State Circuits Recipients biographies*. IEEE. . Retrieved July 2, 2011.

[3] "Network processor startup SiByte raises $40 million in third funding round" (http://www.eetimes.com/electronics-news/4109702/Network-processor-startup-SiByte-raises-40-million-in-third-funding-round). *News release*. May 16, 2000. . Retrieved July 2, 2011.

[4] Loring Wirbel (November 6, 2000). "Broadcom acquires MIPS core provider SiByte" (http://www.eetimes.com/electronics-news/4041235/Broadcom-acquires-MIPS-core-provider-SiByte). *EE Times*. . Retrieved July 2, 2011.

[5] "EDS Members Named Winners of the 2003 IEEE Technical Field Awards" (http://www.ieee.org/organizations/pubs/newsletters/eds/apr03/field_award.html). *EDS Newsletter* (IEEE). April 2003. . Retrieved July 2, 2011.

[6] "IEEE Donald O. Pederson Award in Solid-State Circuits Receipients" (http://www.ieee.org/documents/pederson_rl.pdf). IEEE. . Retrieved July 2, 2011.

[7] "P.A. Semi CEO Dobberpuhl Elected to the National Academy of Engineering" (http://www.hcp.com/news/newsdetails.php/id/7926). *News release* (Highland Capital Partners). February 16, 2006. .

[8] "Distinguished Alumni Award (2003)" (http://www.ece.illinois.edu/alumni/awards/distinguished/03-dobberpuhl.html). University of Illinois. 2003. . Retrieved July 2, 2011.

[9] Tom Moone (May 20, 2009). "Alumnus Dobberpuhl receives College of Engineering Alumni Honor Award" (http://www.ece.illinois.edu/mediacenter/article.asp?id=213). University of Illinois. . Retrieved July 2, 2011.

[10] Katie Marsal (March 13, 2010). "Former P.A. Semi chief leaves Apple for chip startup - report" (http://www.appleinsider.com/articles/10/03/13/former_p_a_semi_chief_leaves_apple_for_chip_startup_report.html). *Apple Insider*. . Retrieved July 2, 2011.

[11] http://www.nap.edu/openbook.php?record_id=12980&page=162

Electrical_engineering

Electrical engineering is a field of engineering that generally deals with the study and application of electricity, electronics and electromagnetism. The field first became an identifiable occupation in the late nineteenth century after commercialization of the electric telegraph and electrical power supply. It now covers a range of subtopics including power, electronics, control systems, signal processing and telecommunications.

Electrical engineers design complex power systems...

Electrical engineering may include electronic engineering. Where a distinction is made, usually outside of the United States, electrical engineering is considered to deal with the problems associated with large-scale electrical systems such as power transmission and motor control, whereas electronic engineering deals with the study of small-scale electronic systems including computers and integrated circuits.[1] Alternatively, electrical engineers are usually concerned with using electricity to transmit energy, while electronic engineers are concerned with using electricity to process information. More recently, the distinction has become blurred by the growth of power electronics.

...and electronic circuits.

History

Electricity has been a subject of scientific interest since at least the early 17th century. The first electrical engineer was probably William Gilbert who designed the versorium: a device that detected the presence of statically charged objects. He was also the first to draw a clear distinction between magnetism and static electricity and is credited with establishing the term electricity.[2] In 1775 Alessandro Volta's scientific experimentations devised the electrophorus, a device that produced a static electric charge, and by 1800 Volta developed the voltaic pile, a forerunner of the electric battery.[3]

The discoveries of Michael Faraday formed the foundation of electric motor technology.

However, it was not until the 19th century that research into the subject started to intensify. Notable developments in this century include the work of Georg Ohm, who in 1827 quantified the relationship between the electric current and potential difference in a conductor, Michael Faraday, the discoverer of electromagnetic induction in 1831, and James Clerk Maxwell, who in 1873 published a unified theory of electricity and magnetism in his treatise *Electricity and Magnetism*.[4]

From the 1830s, efforts were made to apply electricity to practical use in telegraphy. By the end of the 19th century the world had been forever changed by the rapid communication made possible by engineering development of land-line, underwater and, eventually, wireless telegraphy.

Practical applications and advances in such fields created an increasing need for standardized units of measure; it led to the international standardization of the units ohm, volt, ampere, coulomb, and watt. This was achieved at an international conference in Chicago 1893.[5] The publication of these standards formed the basis of future advances in standardisation in various industries, and in many countries the definitions were immediately recognised in relevant legislation.[6]

Thomas Edison built the world's first large-scale electrical supply network.

During these years, the study of electricity was largely considered to be a subfield of physics. It was not until the late 19th century that universities started to offer degrees in electrical engineering. The Darmstadt University of Technology founded the first chair and the first faculty of electrical engineering worldwide in 1882. In the same year, under Professor Charles Cross, the Massachusetts Institute of Technology began offering the first option of Electrical Engineering within a physics department.[7] In 1883 Darmstadt University of Technology and Cornell University introduced the world's first courses of study in electrical engineering, and in 1885 the University College London founded the first chair of electrical engineering in the United Kingdom.[8] The University of Missouri subsequently established the first department of electrical engineering in the United States in 1886.[9]

Nikola Tesla made long-distance electrical transmission networks possible.

During this period, the work concerning electrical engineering increased dramatically. In 1882, Edison switched on the world's first large-scale electrical supply network that provided 110 volts direct current to fifty-nine customers in lower Manhattan. In 1884 Sir Charles Parsons invented the steam turbine which today generates about 80 percent of the electric power in the world using a variety of heat sources. In 1887, Nikola Tesla filed a number of patents related to a competing form of power distribution known as alternating current. In the following years a bitter rivalry between Tesla and Edison, known as the "War of Currents", took place over the preferred method of distribution. AC eventually replaced DC for generation and power distribution, enormously extending the range and improving the safety and efficiency of power distribution.

The efforts of the two did much to further electrical engineering—Tesla's work on induction motors and polyphase systems influenced the field for years to come, while Edison's work on telegraphy and his development of the stock ticker proved lucrative for his company, which ultimately became General Electric. However, by the end of the 19th century, other key figures in the progress of electrical engineering were beginning to emerge.[10]

Modern developments

During the development of radio, many scientists and inventors contributed to radio technology and electronics. In his classic UHF experiments of 1888, Heinrich Hertz transmitted (via a spark-gap transmitter) and detected radio waves using electrical equipment. In 1895, Nikola Tesla was able to detect signals from the transmissions of his New York lab at West Point (a distance of 80.4 km / 49.95 miles).[11] In 1897, Karl Ferdinand Braun introduced the cathode ray tube as part of an oscilloscope, a crucial enabling technology for electronic television.[12] John Fleming invented the first radio tube, the diode, in 1904. Two years later, Robert von Lieben and Lee De Forest independently developed the amplifier tube, called the triode.[13] In 1895, Guglielmo Marconi furthered the art of hertzian wireless methods. Early on, he sent wireless signals over a distance of one and a half miles. In December 1901, he sent wireless waves that were not affected by the curvature of the Earth. Marconi later transmitted the wireless signals across the Atlantic between Poldhu, Cornwall, and St. John's, Newfoundland, a distance of 2100 miles (**unknown operator: u'strong'** km).[14] In 1920 Albert Hull developed the magnetron which would eventually lead to the development of the microwave oven in 1946 by Percy Spencer.[15] [16] In 1934 the British military began to make strides toward radar (which also uses the magnetron) under the direction of Dr Wimperis, culminating in the operation of the first radar station at Bawdsey in August 1936.[17]

In 1941 Konrad Zuse presented the Z3, the world's first fully functional and programmable computer.[18] In 1946 the ENIAC (Electronic Numerical Integrator and Computer) of John Presper Eckert and John Mauchly followed, beginning the computing era. The arithmetic performance of these machines allowed engineers to develop completely new technologies and achieve new objectives, including the Apollo missions and the NASA moon landing.[19]

Jack Kilby's original integrated circuit

The invention of the transistor in 1947 by William B. Shockley, John Bardeen and Walter Brattain opened the door for more compact devices and led to the development of the integrated circuit in 1958 by Jack Kilby and independently in 1959 by Robert Noyce.[20] Starting in 1968, Ted Hoff and a team at Intel invented the first commercial microprocessor, which presaged the personal computer. The Intel 4004 was a 4-bit processor released in 1971, but in 1973 the Intel 8080, an 8-bit processor, made the first personal computer, the Altair 8800, possible.[21]

Education

Electrical engineers typically possess an academic degree with a major in *electrical engineering*, *electronics engineering*, or *electrical and electronic engineering*. The same fundamental principles are taught in all programs, though emphasis may vary according to title. The length of study for such a degree is usually four or five years and the completed degree may be designated as a Bachelor of Engineering, Bachelor of Science, Bachelor of Technology or Bachelor of Applied Science depending upon the university. The degree generally includes units covering physics, mathematics, computer science, project management and specific topics in electrical engineering. Initially such topics cover most, if not all, of the sub-disciplines of electrical engineering. Students then choose to specialize in one or more sub-disciplines towards the end of the degree. In many institutions electronic engineering is included as part of an electrical award, sometimes explicitly (such as a [Bachelor of Engineering] (Electrical and Electronic), in others electrical and electronic engineering are considered sufficiently broad and complex to be considered separately.[22]

Some electrical engineers choose to pursue a postgraduate degree such as a Master of Engineering/Master of Science (M.Eng./M.Sc.), a Master of Engineering Management, a Doctor of Philosophy (Ph.D.) in Engineering, an Engineering Doctorate (Eng.D.), or an Engineer's degree. The Master and Engineer's degree may consist of either research, coursework or a mixture of the two. The Doctor of Philosophy and Engineering Doctorate degrees consist of a significant research component and are often viewed as the entry point to academia. In the United Kingdom and various other European countries, the Master of Engineering is often considered an undergraduate degree of slightly longer duration than the Bachelor of Engineering.[23]

Practicing engineers

In most countries, a Bachelor's degree in engineering represents the first step towards professional certification and the degree program itself is certified by a professional body. After completing a certified degree program the engineer must satisfy a range of requirements (including work experience requirements) before being certified. Once certified the engineer is designated the title of Professional Engineer (in the United States, Canada and South Africa), Chartered Engineer or Incorporated Engineer (in India, Pakistan, the United Kingdom, Ireland and Zimbabwe), Chartered Professional Engineer (in Australia and New Zealand) or European Engineer (in much of the European Union).

The advantages of certification vary depending upon location. For example, in the United States and Canada "only a licensed engineer may seal engineering work for public and private clients".[24] This requirement is enforced by state and provincial legislation such as Quebec's Engineers Act.[25] In other countries, no such legislation exists. Practically all certifying bodies maintain a code of ethics that they expect all members to abide by or risk expulsion.[26] In this way these organizations play an important role in maintaining ethical standards for the profession. Even in jurisdictions where certification has little or no legal bearing on work, engineers are subject to contract law. In cases where an engineer's work fails he or she may be subject to the tort of negligence and, in extreme cases, the charge of criminal negligence. An engineer's work must also comply with numerous other rules and regulations such as building codes and legislation pertaining to environmental law.

Professional bodies of note for electrical engineers include the Institute of Electrical and Electronics Engineers (IEEE) and the Institution of Engineering and Technology (IET). The IEEE claims to produce 30% of the world's literature in electrical engineering, has over 360,000 members worldwide and holds over 3,000 conferences annually.[27] The IET publishes 21 journals, has a worldwide membership of over 150,000, and claims to be the largest professional engineering society in Europe.[28] [29] Obsolescence of technical skills is a serious concern for electrical engineers. Membership and participation in technical societies, regular reviews of periodicals in the field and a habit of continued learning are therefore essential to maintaining proficiency. MIET(Member of the Institution of Engineering and Technology) is recognised in Europe as Electrical and computer (technology) engineer [30]

In Australia, Canada and the United States electrical engineers make up around 0.25% of the labor force (see note). Outside of Europe and North America, engineering graduates per-capita, and hence probably electrical engineering graduates also, are most numerous in Taiwan, Japan, and South Korea.[31]

Tools and work

From the Global Positioning System to electric power generation, electrical engineers have contributed to the development of a wide range of technologies. They design, develop, test and supervise the deployment of electrical systems and electronic devices. For example, they may work on the design of telecommunication systems, the operation of electric power stations, the lighting and wiring of buildings, the design of household appliances or the electrical control of industrial machinery.[32]

Satellite communications is one of many projects an electrical engineer might work on.

Fundamental to the discipline are the sciences of physics and mathematics as these help to obtain both a qualitative and quantitative description of how such systems will work. Today most engineering work involves the use of computers and it is commonplace to use computer-aided design programs when designing electrical systems. Nevertheless, the ability to sketch ideas is still invaluable for quickly communicating with others.

Although most electrical engineers will understand basic circuit theory (that is the interactions of elements such as resistors, capacitors, diodes, transistors and inductors in a circuit), the theories employed by engineers generally depend upon the work they do. For example, quantum mechanics and solid state physics might be relevant to an engineer working on VLSI (the design of integrated circuits), but are largely irrelevant to engineers working with macroscopic electrical systems. Even circuit theory may not be relevant to a person designing telecommunication systems that use off-the-shelf components. Perhaps the most important technical skills for electrical engineers are reflected in university programs, which emphasize strong numerical skills, computer literacy and the ability to understand the technical language and concepts that relate to electrical engineering.

For many engineers, technical work accounts for only a fraction of the work they do. A lot of time may also be spent on tasks such as discussing proposals with clients, preparing budgets and determining project schedules.[33] Many senior engineers manage a team of technicians or other engineers and for this reason project management skills are important. Most engineering projects involve some form of documentation and strong written communication skills are therefore very important.

The workplaces of electrical engineers are just as varied as the types of work they do. Electrical engineers may be found in the pristine lab environment of a fabrication plant, the offices of a consulting firm or on site at a mine. During their working life, electrical engineers may find themselves supervising a wide range of individuals including scientists, electricians, computer programmers and other engineers.

Sub-disciplines

Electrical engineering has many sub-disciplines, the most popular of which are listed below. Although there are electrical engineers who focus exclusively on one of these sub-disciplines, many deal with a combination of them. Sometimes certain fields, such as electronic engineering and computer engineering, are considered separate disciplines in their own right.

Power

Power pole

Power engineering deals with the generation, transmission and distribution of electricity as well as the design of a range of related devices. These include transformers, electric generators, electric motors, high voltage engineering and power electronics. In many regions of the world, governments maintain an electrical network called a power grid that connects a variety of generators together with users of their energy. Users purchase electrical energy from the grid, avoiding the costly exercise of having to generate their own. Power engineers may work on the design and maintenance of the power grid as well as the power systems that connect to it. Such systems are called *on-grid* power systems and may supply the grid with additional power, draw power from the grid or do both. Power engineers may also work on systems that do not connect to the grid, called *off-grid* power systems, which in some cases are preferable to on-grid systems. The future includes Satellite controlled power systems, with feedback in real time to prevent power surges and prevent blackouts.

Control

Control engineering focuses on the modeling of a diverse range of dynamic systems and the design of controllers that will cause these systems to behave in the desired manner. To implement such controllers electrical engineers may use electrical circuits, digital signal processors, microcontrollers and PLCs (Programmable Logic Controllers). Control engineering has a wide range of applications from the flight and propulsion systems of commercial airliners to the cruise control present in many modern automobiles. It also plays an important role in industrial automation.

Control systems play a critical role in space flight.

Control engineers often utilize feedback when designing control systems. For example, in an automobile with cruise control the vehicle's speed is continuously monitored and fed back to the system which adjusts the motor's power output accordingly. Where there is regular feedback, control theory can be used to determine how the system responds to such feedback.

Electronics

Electronic engineering involves the design and testing of electronic circuits that use the properties of components such as resistors, capacitors, inductors, diodes and transistors to achieve a particular functionality. The tuned circuit, which allows the user of a radio to filter out all but a single station, is just one example of such a circuit. Another example (of a pneumatic signal conditioner) is shown in the adjacent photograph.

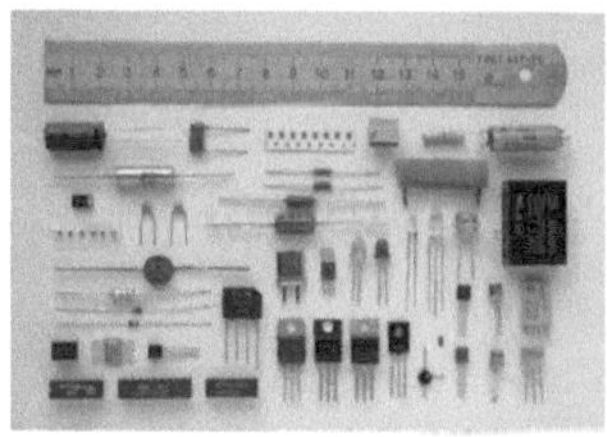
Electronic components

Prior to the second world war, the subject was commonly known as *radio engineering* and basically was restricted to aspects of communications and radar, commercial radio and early television. Later, in post war years, as consumer devices began to be developed, the field grew to include modern television, audio systems, computers and microprocessors. In the mid-to-late 1950s, the term *radio engineering* gradually gave way to the name *electronic engineering*.

Before the invention of the integrated circuit in 1959, electronic circuits were constructed from discrete components that could be manipulated by humans. These discrete circuits consumed much space and power and were limited in speed, although they are still common in some applications. By contrast, integrated circuits packed a large number—often millions—of tiny electrical components, mainly transistors, into a small chip around the size of a coin. This allowed for the powerful computers and other electronic devices we see today.

Microelectronics

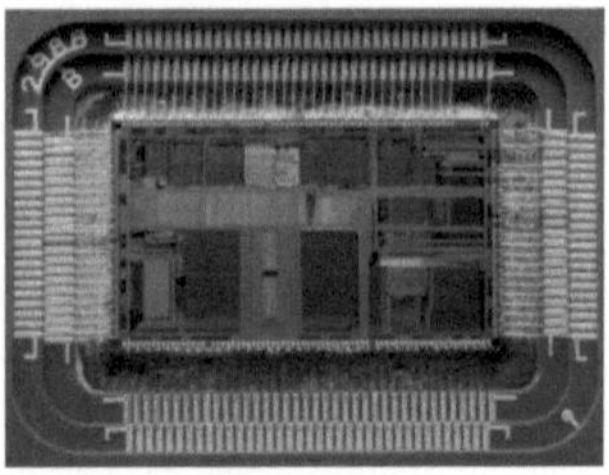

Microprocessor

Microelectronics engineering deals with the design and microfabrication of very small electronic circuit components for use in an integrated circuit or sometimes for use on their own as a general electronic component. The most common microelectronic components are semiconductor transistors, although all main electronic components (resistors, capacitors, inductors) can be created at a microscopic level. Nanoelectronics is the further scaling of devices down to nanometer levels.

Microelectronic components are created by chemically fabricating wafers of semiconductors such as silicon (at higher frequencies, compound semiconductors like gallium arsenide and indium phosphide) to obtain the desired transport of electronic charge and control of current. The field of microelectronics involves a significant amount of chemistry and material science and requires the electronic engineer working in the field to have a very good working knowledge of the effects of quantum mechanics.

Signal processing

A Bayer filter on a CCD requires signal processing to get a red, green, and blue value at each pixel.

Signal processing deals with the analysis and manipulation of signals. Signals can be either analog, in which case the signal varies continuously according to the information, or digital, in which case the signal varies according to a series of discrete values representing the information. For analog signals, signal processing may involve the amplification and filtering of audio signals for audio equipment or the modulation and demodulation of signals for telecommunications. For digital signals, signal processing may involve the compression, error detection and error correction of digitally sampled signals.

Signal Processing is a very mathematically oriented and intensive area forming the core of digital signal processing and it is rapidly expanding with new applications in every field of electrical engineering such as communications, control, radar, TV/Audio/Video engineering, power electronics and bio-medical engineering as many already existing analog systems are replaced with their digital counterparts. Analog signal processing is still important in the design of many control systems.

DSP processor ICs are found in every type of modern electronic systems and products including, SDTV | HDTV sets, radios and mobile communication devices, Hi-Fi audio equipments, Dolby noise reduction algorithms, GSM mobile phones, mp3 multimedia players, camcorders and digital cameras, automobile control systems, noise cancelling headphones, digital spectrum analyzers, intelligent missile guidance, radar, GPS based cruise control systems and all kinds of image processing, video processing, audio processing and speech processing systems.

Telecommunications

Satellite dishes are a crucial component in the analysis of satellite information.

Telecommunications engineering focuses on the transmission of information across a channel such as a coax cable, optical fiber or free space. Transmissions across free space require information to be encoded in a carrier wave to shift the information to a carrier frequency suitable for transmission, this is known as modulation. Popular analog modulation techniques include amplitude modulation and frequency modulation. The choice of modulation affects the cost and performance of a system and these two factors must be balanced carefully by the engineer.

Once the transmission characteristics of a system are determined, telecommunication engineers design the transmitters and receivers needed for such systems. These two are sometimes combined to form a two-way communication device known as a transceiver. A key consideration in the design of transmitters is their power consumption as this is closely related to their signal strength. If the signal strength of a transmitter is insufficient the signal's information will be corrupted by noise.

Instrumentation

Flight instruments provide pilots the tools to control aircraft analytically.

Instrumentation engineering deals with the design of devices to measure physical quantities such as pressure, flow and temperature. The design of such instrumentation requires a good understanding of physics that often extends beyond electromagnetic theory. For example, flight instruments measure variables such as wind speed and altitude to enable pilots the control of aircraft analytically. Similarly, thermocouples use the Peltier-Seebeck effect to measure the temperature difference between two points.

Often instrumentation is not used by itself, but instead as the sensors of larger electrical systems. For example, a thermocouple might be used to help ensure a furnace's temperature remains constant. For this reason, instrumentation engineering is often viewed as the counterpart of control engineering.

Computers

Supercomputers are used in fields as diverse as computational biology and geographic information systems.

Computer engineering deals with the design of computers and computer systems. This may involve the design of new hardware, the design of PDAs and supercomputers or the use of computers to control an industrial plant. Computer engineers may also work on a system's software. However, the design of complex software systems is often the domain of software engineering, which is usually considered a separate discipline. Desktop computers represent a tiny fraction of the devices a computer engineer might work on, as computer-like architectures are now found in a range of devices including video game consoles and DVD players.

Related disciplines

Mechatronics is an engineering discipline which deals with the convergence of electrical and mechanical systems. Such combined systems are known as electromechanical systems and have widespread adoption. Examples include automated manufacturing systems, heating, ventilation and air-conditioning systems and various subsystems of aircraft and automobiles.

The term *mechatronics* is typically used to refer to macroscopic systems but futurists have predicted the emergence of very small electromechanical devices. Already such small devices, known as Microelectromechanical systems (MEMS), are used in automobiles to tell airbags when to deploy, in digital projectors to create sharper images and in inkjet printers to create nozzles for high definition printing. In the future it is hoped the devices will help build tiny implantable medical devices and improve optical communication.[34]

Biomedical engineering is another related discipline, concerned with the design of medical equipment. This includes fixed equipment such as ventilators, MRI scanners and electrocardiograph monitors as well as mobile equipment such as cochlear implants, artificial pacemakers and artificial hearts.

See also

- Analog signal processing
- Computer engineering
- Electronic design automation
- Electronic engineering
- Electrical Technologist
- IEEE
- Institution of Engineering and Technology (IET)
- International Electrotechnical Commission (IEC)
- List of electrical engineering topics (alphabetical)
- List of electrical engineering topics (thematic)
- List of electrical engineers
- List of Russian electrical engineers
- Timeline of electrical and electronic engineering
- Occupations in Electrical/Electronics Engineering

Note

Note I - There were around 300,000 people (as of 2006) working as electrical engineers in the US; in Australia, there were around 17,000 (as of 2008) and in Canada, there were around 37,000 (as of 2007), constituting about 0.2% of the labour force in each of the three countries. Australia and Canada reported that 96% and 88% of their electrical engineers respectively are male.[35]

References

[1] "What is the difference between electrical and electronic engineering?" (http://www.ieee.org/portal/site/mainsite/menuitem.818c0c39e85ef176fb2275875bac26c8/index.jsp?&pName=corp_level1&path=education/faqs&file=faqs1.xml&xsl=generic.xsl). *FAQs - Studying Electrical Engineering*. . Retrieved 20 March 2012.

[2] "William Gilbert (1544–1603)" (http://www.magnet.fsu.edu/education/tutorials/pioneers/gilbert.html). *Pioneers in Electricity*. . Retrieved 13 May 2007.

[3] Vaunt Design Group. (2005). *Inventor Alessandro Volta Biography.* (http://www.ideafinder.com/history/inventors/volta.htm) Troy MI: The Great Idea Finder. Accessed 21 March 2008.

[4] ""Ohm, Georg Simon", "Faraday, Michael" and "Maxwell, James Clerk"". *Encyclopædia Britannica* (11 ed.). 1911.

[5] Proceedings of the International Electrical Congress held in the city of Chicago, August 21st to 25th, 1893. Publisher: New York, American Institute of Electrical Engineers, 1894. May be downloaded from http://www.archive.org/details/proceedingsinte01chicgoog

[6] Tunbridge, Paul. Lord Kelvin: His Influence on Electrical Measurements and Units. Publisher: The Institution of Engineering and Technology1991. ISBN 978-0863412370

[7] Weber, Ernst; Frederik Nebeker (1994). *The Evolution of Electrical Engineering: A Personal Perspective*. IEEE Press. ISBN 0-7803-1066-7.

[8] "Welcome to ECE!" (http://www.ece.cornell.edu). *Cornell University - School of Electrical and Computer Engineering*. . Retrieved 29 December 2005.

[9] Ryder, John; Donald G. Fink (1984). *Engineers and Electrons*. IEEE Press. ISBN 0-87942-172-X.

[10] "History" (http://www.nfpa.org/itemDetail.asp?categoryID=500&itemID=18020&URL=About Us/History). *National Fire Protection Association (NFPA)*. . Retrieved 19 January 2006. *(published 1996 in the NFPA Journal)*

[11] Leland Anderson, "*Nikola Tesla On His Work With Alternating Currents and Their Application to Wireless Telegraphy, Telephony, and Transmission of Power*", Sun Publishing Company, LC 92-60482, ISBN 0-9632652-0-2 (*ed.* excerpts available online (http://www.tfcbooks.com/tesla/nt_on_ac.htm))

[12] "Karl Ferdinand Braun" (http://nobelprize.org/nobel_prizes/physics/laureates/1909/braun-bio.html). . Retrieved 10 September 2006.

[13] "History of Amateur Radio" (http://www.amateurradio.uni-halle.de/hamradio.en.html). *What is Amateur Radio?*. . Retrieved 18 January 2006.

[14] Marconi's biography at Nobelprize.org (http://nobelprize.org/nobel_prizes/physics/laureates/1909/marconi-bio.html) retrieved 21 June 2008.

[15] "Albert W. Hull (1880–1966)" (http://www.ieee.org/organizations/history_center/legacies/hull.html). *IEEE History Center*. . Retrieved 22 January 2006.

[16] "Who Invented Microwaves?" (http://www.gallawa.com/microtech/history.html). . Retrieved 22 January 2006.

[17] "Early Radar History" (http://www.penleyradararchives.org.uk/history/introduction.htm). *Peneley Radar Archives*. . Retrieved 22 January 2006.

[18] "The Z3" (http://irb.cs.tu-berlin.de/~zuse/Konrad_Zuse/en/Rechner_Z3.html). . Retrieved 18 January 2006.

[19] "The ENIAC Museum Online" (http://www.seas.upenn.edu/~museum/guys.html). . Retrieved 18 January 2006.

[20] "Electronics Timeline" (http://www.greatachievements.org/?id=3956). *Greatest Engineering Achievements of the Twentieth Century*. . Retrieved 18 January 2006.

[21] "Computing History (1971–1975)" (http://mbinfo.mbdesign.net/1971-75.htm). . Retrieved 18 January 2006.

[22] "Electrical and Electronic Engineering" (http://www.flinders.edu.au/science_engineering/csem/disciplines/eee/). . Retrieved 8 December 2011.

[23] Various including graduate degree requirements at MIT (http://www.eecs.mit.edu/grad/degrees.html), study guide at UWA (http://www.ecm.uwa.edu.au/students/study-guides-2012/be-elec-mech), the curriculum at Queen's (http://www.queensu.ca/calendars/appsci/pg219.html) and unit tables at Aberdeen (http://www.abdn.ac.uk/registry/calendar/requirements/07H50116.doc)

[24] "Why Should You Get Licensed?" (http://web.archive.org/web/20050604085233/http://www.nspe.org/lc1-why.asp). *National Society of Professional Engineers*. Archived from the original (http://www.nspe.org/lc1-why.asp) on June 4, 2005. . Retrieved 11 July 2005.

[25] "Engineers Act" (http://www2.publicationsduquebec.gouv.qc.ca/dynamicSearch/telecharge.php?type=2&file=//I_9/I9_A.htm). *Quebec Statutes and Regulations (CanLII)*. . Retrieved 24 July 2005.

[26] "Codes of Ethics and Conduct" (http://onlineethics.org/CMS/profpractice/ethcodes.aspx). *Online Ethics Center*. . Retrieved 24 July 2005.

[27] "About the IEEE" (http://www.ieee.org/about/). *IEEE*. . Retrieved 11 July 2005.

[28] "About the IET" (http://www.theiet.org/about/). *The IET*. . Retrieved 11 July 2005.

[29] "Journal and Magazines" (http://www.theiet.org/publishing/journals/). *The IET*. . Retrieved 11 July 2005.

[30] "Electrical and Electronics Engineers, except Computer" (http://web.archive.org/web/20050713014728/http://www.bls.gov/oco/ocos031.htm). *Occupational Outlook Handbook*. Archived from the original (http://www.bls.gov/oco/ocos031.htm) on July 13, 2005. . Retrieved 16 July 2005.. (see here regarding copyright)

[31] "Science and Engineering Indicators 2004, Appendix 2-33" (http://www.nsf.gov/statistics/seind04/append/c2/at02-33.pdf) (PDF). National Science Foundation. 2004. .

[32] "Electrical and Electronics Engineers, except Computer" (http://web.archive.org/web/20050713014728/http://www.bls.gov/oco/ocos031.htm). *Occupational Outlook Handbook*. Archived from the original (http://www.bls.gov/oco/ocos031.htm) on 13 July 2005. . Retrieved 16 July 2005. (see Internet Archive (http://web.archive.org/web/www.bls.gov/oco/ocos031.htm))

[33] Trevelyan, James; (2005). *What Do Engineers Really Do?*. University of Western Australia. (seminar with slides (http://www.mech.uwa.edu.au/jpt/Engineering Roles 050503.pdf))

[34] "MEMS the world!" (http://web.archive.org/web/20050317085002/http://www.intellisensesoftware.com/Technology.html). *IntelliSense Software Corporation*. Archived from the original (http://www.intellisensesoftware.com/Technology.html) on March 17, 2005. . Retrieved 17 July 2005.

[35] "Electrical Engineers" (http://www.bls.gov/oco/ocos027.htm). Bureau of Labor Statistics. . Retrieved 13 March 2009. See also: "Work Experience of the Population in 2006" (http://www.bls.gov/news.release/History/work_12192007.txt). Bureau of Labor Statistics. . Retrieved 20 June 2008. and "Electrical and Electronics Engineers" (http://joboutlook.gov.au/Pages/occupation.aspx?search=alpha&tab=prospects&cluster=&code=2333). *Australian Careers*. . Retrieved 13 March 2009. and "Electrical and Electronics Engineers" (http://www.jobfutures.ca/noc/2133p1.shtml). Canadian jobs service. . Retrieved 13 March 2009.

External links

- IEEE Global History Network (http://www.ieeeghn.org/) A wiki-based site with many resources about the history of IEEE, its members, their professions and electrical and informational technologies and sciences.
- International Electrotechnical Commission (IEC) (http://www.iec.ch/)
- MIT OpenCourseWare (http://ocw.mit.edu/OcwWeb/Electrical-Engineering-and-Computer-Science/index.htm) In-depth look at Electrical Engineering with online courses featuring video lectures.

Microprocessor

A **microprocessor** incorporates the functions of a computer's central processing unit (CPU) on a single integrated circuit,[1] (IC) or at most a few integrated circuits.[2] It is a multipurpose, programmable device that accepts digital data as input, processes it according to instructions stored in its memory, and provides results as output. It is an example of sequential digital logic, as it has internal memory. Microprocessors operate on numbers and symbols represented in the binary numeral system.

Intel 4004, the first general-purpose, commercial microprocessor

The advent of low-cost computers on integrated circuits has transformed modern society. General-purpose microprocessors in personal computers are used for computation, text editing, multimedia display, and communication over the Internet. Many more microprocessors are part of embedded systems, providing digital control of a myriad of objects from appliances to automobiles to cellular phones and industrial process control.

Origins

During the 1960s, computer processors were constructed out of small and medium-scale ICs each containing from tens to a few hundred transistors. For each computer built, all of these had to be placed and soldered onto printed circuit boards, and often multiple boards would have to be interconnected in a chassis. The large number of discrete logic gates used more electrical power—and therefore, produced more heat—than a more integrated design with fewer ICs. The distance that signals had to travel between ICs on the boards limited the speed at which a computer could operate.

In the NASA Apollo space missions to the moon in the 1960s and 1970s, all onboard computations for primary guidance, navigation and control were provided by a small custom processor called "The Apollo Guidance Computer". It used a primitive gate array whose only logic elements were three-input NOR gates.[3]

The integration of a whole CPU onto a single chip or on a few chips greatly reduced the cost of processing power. The integrated circuit processor was produced in large numbers by highly automated processes, so unit cost was low. Single-chip processors increase reliability as there were many fewer electrical connections to fail. As microprocessor designs get faster, the cost of manufacturing a chip (with smaller components built on a semiconductor chip the same size) generally stays the same.

Microprocessors integrated into one or a few large-scale ICs the architectures that had previously been implemented using many medium- and small-scale integrated circuits. Continued increases in microprocessor capacity have rendered other forms of computers almost completely obsolete (see history of computing hardware), with one or more microprocessors used in everything from the smallest embedded systems and handheld devices to the largest mainframes and supercomputers.

The first microprocessors emerged in the early 1970s and were used for electronic calculators, using binary-coded decimal (BCD) arithmetic on 4-bit words. Other embedded uses of 4-bit and 8-bit microprocessors, such as terminals, printers, various kinds of automation etc., followed soon after. Affordable 8-bit microprocessors with 16-bit addressing also led to the first general-purpose microcomputers from the mid-1970s on.

Since the early 1970s, the increase in capacity of microprocessors has followed Moore's law; this originally suggested that the number of transistors that can be fitted onto a chip doubles every year,[4] though Moore later refined the period to two years.[5]

Embedded applications

Thousands of items that were traditionally not computer-related include microprocessors. These include large and small household appliances, cars (and their accessory equipment units), car keys, tools and test instruments, toys, light switches/dimmers and electrical circuit breakers, smoke alarms, battery packs, and hi-fi audio/visual components (from DVD players to phonograph turntables.) Such products as cellular telephones, DVD video system and ATSC HDTV broadcast system fundamentally require consumer devices with powerful, low-cost, microprocessors. Increasingly stringent pollution control standards effectively require automobile manufacturers to use microprocessor engine management systems, to allow optimal control of emissions over widely varying operating conditions of an automobile. Non-programmable controls would require complex, bulky, or costly implementation to achieve the results possible with a microprocessor.

A microprocessor control program can be easily tailored to different needs of a product line, allowing upgrades in performance with minimal redesign of the product. Different features can be implemented in different models of a product line at negligible production cost.

Microprocessor control of a system can provide control strategies that would be impractical to implement using electromechanical controls or purpose-built electronic controls. For example, an engine control system in an automobile can adjust ignition timing based on engine speed, load on the engine, ambient temperature, and any observed tendency for knocking - allowing an automobile to operate on a range of fuel grades.

Structure

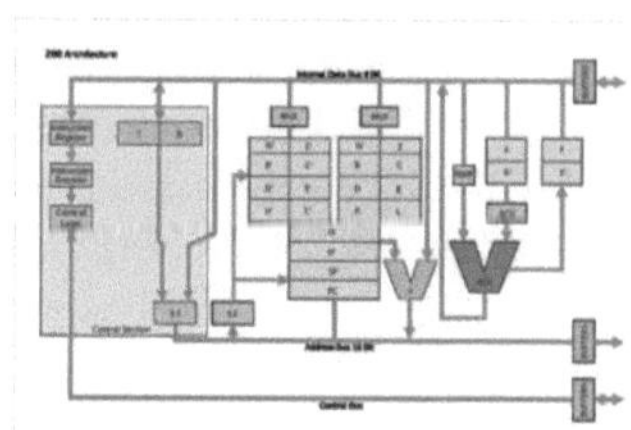
A block diagram of the internal architecture of the Z80 microprocessor, showing the arithmetic and logic section, register file, control logic section, and buffers to external address and data lines

The internal arrangement of a microprocessor varies depending on the age of the design and the intended purposes of the processor. The complexity of an integrated circuit is bounded by physical limitations of the number of transistors that can be put onto one chip, the number of package terminations that can connect the processor to other parts of the system, the number of interconnections it is possible to make on the chip, and the heat that the chip can dissipate. Advancing technology makes more complex and powerful chips feasible to manufacture.

A minimal hypothetical microprocessor might only include an arithmetic logic unit (ALU) and a control logic section. The ALU performs operations such as addition, subtraction, and operations such as AND or OR. Each operation of the ALU sets one or more flags in a status register, which indicate the results of the last operation (zero value, negative number, overflow. or others). The logic section retrieves instruction operation codes from memory, and initiates whatever sequence of operations of the ALU required to carry out the instruction. A single operation code might affect many individual data paths, registers, and other elements of the processor.

As integrated circuit technology advanced, it was feasible to manufacture more and more complex processors on a single chip. The size of data objects became larger; allowing more transistors on a chip allowed word sizes to increase from 4- and 8-bit words up to today's 64-bit words. Additional features were added to the processor architecture; more on-chip registers speeded up programs, and complex instructions could be used to make more compact programs. Floating-point arithmetic, for example, was often not available on 8-bit microprocessors, but had to be carried out in software. Integration of the floating point unit first as a separate integrated circuit and then as part of the same microprocessor chip, speeded up floating point calculations.

Occasionally the physical limitations of integrated circuits made such practices as a bit slice approach necessary. Instead of processing all of a long word on one integrated circuit, multiple circuits in parallel processed subsets of each data word. While this required extra logic to handle, for example, carry and overflow within each slice, the result was a system that could handle, say, 32-bit words using integrated circuits with a capacity for only 4 bits each.

With the ability to put large numbers of transistors on one chip, it becomes feasible to integrate memory on the same die as the processor. This CPU cache has the advantage of faster access than off-chip memory, and increases the processing speed of the system for many applications. Generally, processor speed has increased more rapidly than external memory speed, so cache memory is necessary if the processor is not to be delayed by slower external memory.

Firsts

Three projects delivered a microprocessor at about the same time: Garrett AiResearch's Central Air Data Computer (CADC) (1968), Texas Instruments (TI) TMS 1000 (1971 September), and Intel's 4004 (1971 November).

Intel 4004

The 4004 with cover removed (left) and as actually used (right)

The Intel 4004 is generally regarded as the first commercially available microprocessor,[6] [7] and cost $60.[8] The first known advertisement for the 4004 is dated November 15, 1971 and appeared in Electronic News.[9] The project that produced the 4004 originated in 1969, when Busicom, a Japanese calculator manufacturer, asked Intel to build a chipset for high-performance desktop calculators. Busicom's original design called for a programmable chip set consisting of seven different chips. Three of the chips were to make a special-purpose CPU with its program stored in ROM and its data stored in shift register read-write memory. Ted Hoff, the Intel engineer assigned to evaluate the project, believed the Busicom design could be simplified by using dynamic RAM storage for data, rather than shift register memory, and a more traditional general-purpose CPU architecture. Hoff came up with a four–chip architectural proposal: a ROM chip for storing the programs, a dynamic RAM chip for storing data, a simple I/O device and a 4-bit central processing unit (CPU). Although not a chip designer, he felt the CPU could be integrated into a single chip, but as he lacked the technical know-how the idea remained just a wish for the time being.

While the architecture and specifications of the MCS-4 came from the interaction of Hoff with Stanley Mazor, a software engineer reporting to him, and with Busicom engineer Masatoshi Shima, during 1969, Mazor and Hoff moved on to other projects while in April 1970, Intel hired Federico Faggin as project leader, a move that ultimately made the single-chip CPU final design a reality (Shima instead designed the Busicom calculator firmware and assisted Faggin during the first six months of the implementation). Faggin, who originally developed the silicon gate technology (SGT) in 1968 at Fairchild Semiconductor[10] and designed the world's first commercial integrated circuit using SGT, the Fairchild 3708, had the correct background to lead the project into what would become the first commercial general purpose microprocessor, since it was his very own invention, SGT in addition to his new methodology for random logic design, that made it possible to implement a single-chip CPU with the proper speed, power dissipation and cost. The manager of Intel's MOS Design Department was Leslie L. Vadász. at the time of the MCS-4 development, but Vadasz's attention was completely focused on the mainstream business of semiconductor memories and he left the leadership and the management of the MCS-4 project to Faggin, which was ultimately responsible for leading the 4004 project to its outcome. Production units of the 4004 were first delivered to Busicom in March 1971 and shipped to other customers in late 1971.

TMS 1000

The Smithsonian Institution says TI engineers Gary Boone and Michael Cochran succeeded in creating the first microcontroller (also called a microcomputer) in 1971. The result of their work was the TMS 1000, which went commercial in 1974.[11]

TI developed the 4-bit TMS 1000 and stressed pre-programmed embedded applications, introducing a version called the TMS1802NC on September 17, 1971 which implemented a calculator on a chip.

TI filed for the patent on the microprocessor. Gary Boone was awarded U.S. Patent 3757306 [12] for the single-chip microprocessor architecture on September 4, 1973. It may never be known which company actually had the first working microprocessor running on the lab bench. In both 1971 and 1976, Intel and TI entered into broad patent cross-licensing agreements, with Intel paying royalties to TI for the microprocessor patent. A history of these events is contained in court documentation from a legal dispute between Cyrix and Intel, with TI as intervenor and owner of the microprocessor patent.

A computer-on-a-chip combines the microprocessor core (CPU), memory, and I/O (input/output) lines onto one chip. The computer-on-a-chip patent, called the "microcomputer patent" at the time, U.S. Patent 4074351 [13], was awarded to Gary Boone and Michael J. Cochran of TI. Aside from this patent, the standard meaning of microcomputer is a computer using one or more microprocessors as its CPU(s), while the concept defined in the patent is more akin to a microcontroller.

Pico/General Instrument

In 1971 Pico Electronics[14] and General Instrument (GI) introduced their first collaboration in ICs, a complete single chip calculator IC for the Monroe/Litton Royal Digital III calculator. This chip could also arguably lay claim to be one of the first microprocessors or microcontrollers having ROM, RAM and a RISC instruction set on-chip. The layout for the four layers of the PMOS process was hand drawn at x500 scale on mylar film, a significant task at the time given the complexity of the chip.

The PICO1/GI250 chip introduced in 1971. This was designed by Pico Electronics (Glenrothes, Scotland) and manufactured by General Instrument of Hicksville NY

Pico was a spinout by five GI design engineers whose vision was to create single chip calculator ICs. They had significant previous design experience on multiple calculator chipsets with both GI and Marconi-Elliott.[15] The key team members had originally been tasked by Elliott Automation to create an 8 bit computer in MOS and had helped establish a MOS Research Laboratory in Glenrothes, Scotland in 1967.

Calculators were becoming the largest single market for semiconductors and Pico and GI went on to have significant success in this burgeoning market. GI continued to innovate in microprocessors and microcontrollers with products including the CP1600, IOB1680 and PIC1650.[16] In 1987 the GI Microelectronics business was spun out into the Microchip PIC microcontroller business.

CADC

In 1968, Garrett AiResearch (which employed designers Ray Holt and Steve Geller) was invited to produce a digital computer to compete with electromechanical systems then under development for the main flight control computer in the US Navy's new F-14 Tomcat fighter. The design was complete by 1970, and used a MOS-based chipset as the core CPU. The design was significantly (approximately 20 times) smaller and much more reliable than the mechanical systems it competed against, and was used in all of the early Tomcat models. This system contained "a 20-bit, pipelined, parallel multi-microprocessor". The Navy refused to allow publication of the design until 1997. For this reason the CADC, and the MP944 chipset it used, are fairly unknown.[17] Ray Holt graduated California Polytechnic University in 1968, and began his computer design career with the CADC. From its inception, it was shrouded in secrecy until 1998 when at Holt's request, the US Navy allowed the documents into the public domain. Since then several have debated if this was the first microprocessor. Holt has stated that no one has compared this microprocessor with those that came later.[18] According to Parab et al. (2007), *"The scientific papers and literature published around 1971 reveal that the MP944 digital processor used for the F-14 Tomcat aircraft of the US Navy qualifies as the first microprocessor. Although interesting, it was not a single-chip processor, and was not general purpose – it was more like a set of parallel building blocks you could use to make a special-purpose DSP form. It indicates that today's industry theme of converging DSP-microcontroller architectures was started in 1971."*[19] This convergence of DSP and microcontroller architectures is known as a Digital Signal Controller.

Gilbert Hyatt

Gilbert Hyatt was awarded a patent claiming an invention pre-dating both TI and Intel, describing a "microcontroller".[20] The patent was later invalidated, but not before substantial royalties were paid out.[21] [22]

Four-Phase Systems AL1

The Four-Phase Systems AL1 was an 8-bit bit slice chip containing eight registers and an ALU.[23] It was designed by Lee Boysel in 1969.[24] [25] [26] At the time, it formed part of a nine-chip, 24-bit CPU with three AL1s, but it was later called a microprocessor when, in response to 1990s litigation by Texas Instruments, a demonstration system was constructed where a single AL1 formed part of a courtroom demonstration computer system, together with RAM, ROM, and an input-output device.[27]

8-bit designs

The Intel 4004 was followed in 1972 by the Intel 8008, the world's first 8-bit microprocessor. The 8008 was not, however, an extension of the 4004 design, but instead the culmination of a separate design project at Intel, arising from a contract with Computer Terminals Corporation, of San Antonio TX, for a chip for a terminal they were designing,[28] the Datapoint 2200 — fundamental aspects of the design came not from Intel but from CTC. In 1968, CTC's Austin O. "Gus" Roche developed the original design for the instruction set and operation of the processor. In 1969, CTC contracted two companies, Intel and Texas Instruments, to make a single-chip implementation, known as the CTC 1201.[29] In late 1970 or early 1971, TI dropped out being unable to make a reliable part. In 1970, with Intel yet to deliver the part, CTC opted to use their own implementation in the Datapoint 3300, using traditional TTL logic instead (thus the first machine to run "8008 code" was not in fact a microprocessor at all). Intel's version of the 1201 microprocessor arrived in late 1971, but was too late, slow, and required a number of additional support chips. CTC had no interest in using it. CTC had originally contracted Intel for the chip, and would have owed them $50,000 for their design work.[29] To avoid paying for a chip they did not want (and could not use), CTC released Intel from their contract and allowed them free use of the design.[29] Intel marketed it as the 8008 in April, 1972, as the world's first 8-bit microprocessor. It was the basis for the famous "Mark-8" computer kit advertised in the magazine *Radio-Electronics* in 1974.

The 8008 was the precursor to the very successful Intel 8080 (1974), which offered much improved performance over the 8008 and required fewer support chips, Zilog Z80 (1976), and derivative Intel 8-bit processors. The competing Motorola 6800 was released August 1974 and the similar MOS Technology 6502 in 1975 (both designed largely by the same people). The 6502 family rivaled the Z80 in popularity during the 1980s.

A low overall cost, small packaging, simple computer bus requirements, and sometimes the integration of extra circuitry (e.g. the Z80's built-in memory refresh circuitry) allowed the home computer "revolution" to accelerate sharply in the early 1980s. This delivered such inexpensive machines as the Sinclair ZX-81, which sold for US$99. A variation of the 6502, the MOS Technology 6510 was used in the Commodore 64 and yet another variant, the 8502, powered the Commodore 128.

The Western Design Center, Inc (WDC) introduced the CMOS 65C02 in 1982 and licensed the design to several firms. It was used as the CPU in the Apple IIe and IIc personal computers as well as in medical implantable grade pacemakers and defibrillators, automotive, industrial and consumer devices. WDC pioneered the licensing of microprocessor designs, later followed by ARM and other microprocessor Intellectual Property (IP) providers in the 1990s.

Motorola introduced the MC6809 in 1978, an ambitious and thought-through 8-bit design source compatible with the 6800 and implemented using purely hard-wired logic. (Subsequent 16-bit microprocessors typically used microcode to some extent, as CISC design requirements were getting too complex for purely hard-wired logic only.)

Another early 8-bit microprocessor was the Signetics 2650, which enjoyed a brief surge of interest due to its innovative and powerful instruction set architecture.

A seminal microprocessor in the world of spaceflight was RCA's RCA 1802 (aka CDP1802, RCA COSMAC) (introduced in 1976), which was used onboard the Galileo probe to Jupiter (launched 1989, arrived 1995). RCA COSMAC was the first to implement CMOS technology. The CDP1802 was used because it could be run at very low power, and because a variant was available fabricated using a special production process (Silicon on Sapphire), providing much better protection against cosmic radiation and electrostatic discharges than that of any other processor of the era. Thus, the SOS version of the 1802 was said to be the first radiation-hardened microprocessor.

The RCA 1802 had what is called a *static design*, meaning that the clock frequency could be made arbitrarily low, even to 0 Hz, a total stop condition. This let the Galileo spacecraft use minimum electric power for long uneventful stretches of a voyage. Timers and/or sensors would awaken/improve the performance of the processor in time for important tasks, such as navigation updates, attitude control, data acquisition, and radio communication. Current versions of the Western Design Center 65C02 and 65C816 have static cores and thus will retain data even when the clock is completely halted.

12-bit designs

The Intersil 6100 family consisted of a 12-bit microprocessor (the 6100) and a range of peripheral support and memory ICs. The microprocessor recognised the DEC PDP-8 minicomputer instruction set. As such it was sometimes referred to as the **CMOS-PDP8**. Since it was also produced by Harris Corporation, it was also known as the **Harris HM-6100**. By virtue of its CMOS technology and associated benefits, the 6100 was being incorporated into some military designs until the early 1980s.

16-bit designs

The first multi-chip 16-bit microprocessor was the National Semiconductor IMP-16, introduced in early 1973. An 8-bit version of the chipset was introduced in 1974 as the IMP-8.

Other early multi-chip 16-bit microprocessors include one used by Digital Equipment Corporation (DEC) in the LSI-11 OEM board set and the packaged PDP 11/03 minicomputer, and the Fairchild Semiconductor MicroFlame 9440, both of which were introduced in the 1975 to 1976 timeframe.

In 1975, National introduced the first 16-bit single-chip microprocessor, the National Semiconductor PACE, which was later followed by an NMOS version, the INS8900.

Another early single-chip 16-bit microprocessor was TI's TMS 9900, which was also compatible with their TI-990 line of minicomputers. The 9900 was used in the TI 990/4 minicomputer, the TI-99/4A home computer, and the TM990 line of OEM microcomputer boards. The chip was packaged in a large ceramic 64-pin DIP package, while most 8-bit microprocessors such as the Intel 8080 used the more common, smaller, and less expensive plastic 40-pin DIP. A follow-on chip, the TMS 9980, was designed to compete with the Intel 8080, had the full TI 990 16-bit instruction set, used a plastic 40-pin package, moved data 8 bits at a time, but could only address 16 KB. A third chip, the TMS 9995, was a new design. The family later expanded to include the 99105 and 99110.

The Western Design Center (WDC) introduced the CMOS 65816 16-bit upgrade of the WDC CMOS 65C02 in 1984. The 65816 16-bit microprocessor was the core of the Apple IIgs and later the Super Nintendo Entertainment System, making it one of the most popular 16-bit designs of all time.

Intel "upsized" their 8080 design into the 16-bit Intel 8086, the first member of the x86 family, which powers most modern PC type computers. Intel introduced the 8086 as a cost effective way of porting software from the 8080 lines, and succeeded in winning much business on that premise. The 8088, a version of the 8086 that used an 8-bit external data bus, was the microprocessor in the first IBM PC. Intel then released the 80186 and 80188, the 80286 and, in 1985, the 32-bit 80386, cementing their PC market dominance with the processor family's backwards compatibility. The 80186 and 80188 were essentially versions of the 8086 and 8088, enhanced with some onboard peripherals and a few new instructions; they were not used in IBM-compatible PCs because the built-in perpherals and their locations in the memory map were incompatible with the IBM design. The 8086 and successors had an innovative but limited method of memory segmentation, while the 80286 introduced a full-featured segmented memory management unit (MMU). The 80386 introduced a flat 32-bit memory model with paged memory management.

The Intel x86 processors up to and including the 80386 do not include floating-point units (FPUs). Intel introduced the 8087, 80287, and 80387 math coprocessors to add hardware floating-point and transcendental function capabilities to the 8086 through 80386 CPUs. The 8087 works with the 8086/8088 and 80186/80188,[30] the 80187 works with the 80186/80188, the 80287 works with the 80286 and 80386,[31] and the 80387 works with the 80386 (yielding better performance than the 80287). The combination of an x86 CPU and an x87 coprocessor forms a single multi-chip microprocessor; the two chips are programmed as a unit using a single integrated instruction set.[32] Though the 8087 coprocessor is interfaced to the CPU through I/O ports in the CPU's address space, this is transparent to the program, which does not need to know about or access these I/O ports directly; the program accesses the coprocessor and its registers through normal instruction opcodes. Starting with the successor to the 80386, the 80486, the FPU was integrated with the control unit, MMU, and integer ALU in a pipelined design on a single chip (in the 80486DX version), or the FPU was eliminated entirely (in the 80486SX version). An ostensible coprocessor for the 80486SX, the 80487, was actually a complete 80486DX which disabled and replaced the coprocessorless 80486SX that it was installed to upgrade.

32-bit designs

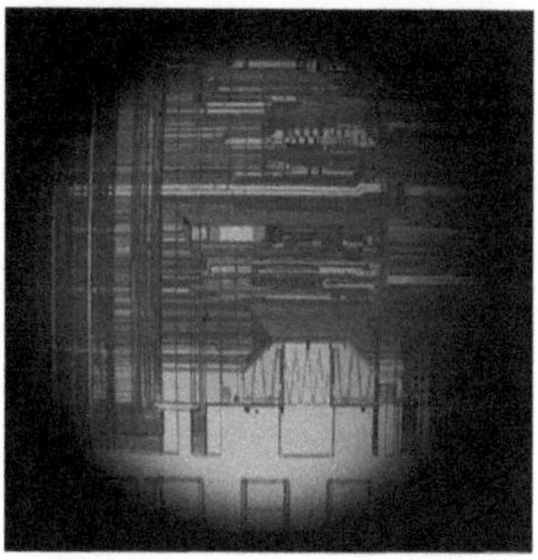
Upper interconnect layers on an Intel 80486DX2 die

16-bit designs had only been on the market briefly when 32-bit implementations started to appear.

The most significant of the 32-bit designs is the MC68000, introduced in 1979. The 68K, as it was widely known, had 32-bit registers in its programming model but used 16-bit internal data paths, 3 16-bit Arithmetic Logic Units, and a 16-bit external data bus (to reduce pin count), and externally supported only 24-bit addresses (internally it worked with full 32 bit addresses). In PC-based IBM-compatible mainframes the MC68000 internal microcode was modified to emulate the 32-bit System/370 IBM mainframe.[33] Motorola generally described it as a 16-bit processor, though it clearly has 32-bit capable architecture. The combination of high performance, large (16 megabytes or 2^{24} bytes) memory space and fairly low cost made it the most popular CPU design of its class. The Apple Lisa and Macintosh designs made use of the 68000, as did a host of other designs in the mid-1980s, including the Atari ST and Commodore Amiga.

The world's first single-chip fully 32-bit microprocessor, with 32-bit data paths, 32-bit buses, and 32-bit addresses, was the AT&T Bell Labs BELLMAC-32A, with first samples in 1980, and general production in 1982[34] [35] After the divestiture of AT&T in 1984, it was renamed the WE 32000 (WE for Western Electric), and had two follow-on generations, the WE 32100 and WE 32200. These microprocessors were used in the AT&T 3B5 and 3B15 minicomputers; in the 3B2, the world's first desktop supermicrocomputer; in the "Companion", the world's first 32-bit laptop computer; and in "Alexander", the world's first book-sized supermicrocomputer, featuring ROM-pack memory cartridges similar to today's gaming consoles. All these systems ran the UNIX System V operating system.

Intel's first 32-bit microprocessor was the iAPX 432, which was introduced in 1981 but was not a commercial success. It had an advanced capability-based object-oriented architecture, but poor performance compared to contemporary architectures such as Intel's own 80286 (introduced 1982), which was almost four times as fast on typical benchmark tests. However, the results for the iAPX432 was partly due to a rushed and therefore suboptimal Ada compiler.

The ARM first appeared in 1985. This is a RISC processor design, which has since come to dominate the 32-bit embedded systems processor space due in large part to its power efficiency, its licensing model, and its wide selection of system development tools. Semiconductor manufacturers generally license cores such as the ARM11 and integrate them into their own system on a chip products; only a few such vendors are licensed to modify the ARM cores. Most cell phones include an ARM processor, as do a wide variety of other products. There are microcontroller-oriented ARM cores without virtual memory support, as well as SMP applications processors with virtual memory.

Motorola's success with the 68000 led to the MC68010, which added virtual memory support. The MC68020, introduced in 1985 added full 32-bit data and address buses. The 68020 became hugely popular in the Unix supermicrocomputer market, and many small companies (e.g., Altos, Charles River Data Systems) produced desktop-size systems. The MC68030 was introduced next, improving upon the previous design by integrating the MMU into the chip. The continued success led to the MC68040, which included an FPU for better math performance. A 68050 failed to achieve its performance goals and was not released, and the follow-up MC68060 was released into a market saturated by much faster RISC designs. The 68K family faded from the desktop in the early 1990s.

Other large companies designed the 68020 and follow-ons into embedded equipment. At one point, there were more 68020s in embedded equipment than there were Intel Pentiums in PCs.[36] The ColdFire processor cores are derivatives of the venerable 68020.

During this time (early to mid-1980s), National Semiconductor introduced a very similar 16-bit pinout, 32-bit internal microprocessor called the NS 16032 (later renamed 32016), the full 32-bit version named the NS 32032. Later the NS 32132 was introduced which allowed two CPUs to reside on the same memory bus, with built in arbitration. The NS32016/32 outperformed the MC68000/10 but the NS32332 which arrived at approximately the same time the MC68020 did not have enough performance. The third generation chip, the NS32532 was different. It had about double the performance of the MC68030 which was released around the same time. The appearance of RISC processors like the AM29000 and MC88000 (now both dead) influenced the architecture of the final core, the NS32764. Technically advanced, using a superscalar RISC core, internally overclocked, with a 64 bit bus, it was still capable of executing Series 32000 instructions through real time translation.

When National Semiconductor decided to leave the Unix market, the chip was redesigned into the Swordfish Embedded processor with a set of on chip peripherals. The chip turned out to be too expensive for the laser printer market and was killed. The design team went to Intel and there designed the Pentium processor, which is very similar to the NS32764 core internally. The big success of the Series 32000 was in the laser printer market, where the NS32CG16 with microcoded BitBlt instructions had very good price/performance and was adopted by large companies like Canon. By the mid-1980s, Sequent introduced the first symmetric multiprocessor (SMP) server-class computer using the NS 32032. This was one of the design's few wins, and it disappeared in the late 1980s. The MIPS R2000 (1984) and R3000 (1989) were highly successful 32-bit RISC microprocessors. They were used in high-end workstations and servers by SGI, among others. Other designs included the interesting Zilog Z80000, which arrived too late to market to stand a chance and disappeared quickly.

In the late 1980s, "microprocessor wars" started killing off some of the microprocessors. Apparently, with only one major design win, Sequent, the NS 32032 just faded out of existence, and Sequent switched to Intel microprocessors.

From 1985 to 2003, the 32-bit x86 architectures became increasingly dominant in desktop, laptop, and server markets, and these microprocessors became faster and more capable. Intel had licensed early versions of the architecture to other companies, but declined to license the Pentium, so AMD and Cyrix built later versions of the architecture based on their own designs. During this span, these processors increased in complexity (transistor count) and capability (instructions/second) by at least three orders of magnitude. Intel's Pentium line is probably the most famous and recognizable 32-bit processor model, at least with the public at large.

64-bit designs in personal computers

While 64-bit microprocessor designs have been in use in several markets since the early 1990s, the early 2000s saw the introduction of 64-bit microprocessors targeted at the PC market.

With AMD's introduction of a 64-bit architecture backwards-compatible with x86, x86-64 (also called **AMD64**), in September 2003, followed by Intel's near fully compatible 64-bit extensions (first called IA-32e or EM64T, later renamed **Intel 64**), the 64-bit desktop era began. Both versions can run 32-bit legacy applications without any performance penalty as well as new 64-bit software. With operating systems Windows XP x64, Windows Vista x64, Windows 7 x64, Linux, BSD and Mac OS X that run 64-bit native, the software is also geared to fully utilize the capabilities of such processors. The move to 64 bits is more than just an increase in register size from the IA-32 as it also doubles the number of general-purpose registers.

The move to 64 bits by PowerPC processors had been intended since the processors' design in the early 90s and was not a major cause of incompatibility. Existing integer registers are extended as are all related data pathways, but, as was the case with IA-32, both floating point and vector units had been operating at or above 64 bits for several years. Unlike what happened when IA-32 was extended to x86-64, no new general purpose registers were added in 64-bit PowerPC, so any performance gained when using the 64-bit mode for applications making no use of the larger address space is minimal.

Multicore designs

A different approach to improving a computer's performance is to add extra processors, as in symmetric multiprocessing designs, which have been popular in servers and workstations since the early 1990s. Keeping up with Moore's Law is becoming increasingly challenging as chip-making technologies approach their physical limits.

In response, microprocessor manufacturers look for other ways to improve performance in order to hold on to the momentum of constant upgrades in the market.

A multi-core processor is simply a single chip containing more than one microprocessor core. This effectively multiplies the processor's potential performance by the number of cores (as long as the operating system and software is designed to take advantage of more than one processor core). Some components, such as bus interface and cache, may be shared between cores. Because the cores are physically very close to each other, they can communicate with each other much faster than separate processors in a multiprocessor system, which improves overall system performance.

In 2005, the first personal computer dual-core processors were announced. As of 2012, dual-core and quad-core processors are widely used in home PCs and laptops while quad, six, eight, ten, twelve, and sixteen-core processors are common in the professional and enterprise markets with workstations and servers.

Sun Microsystems has released the Niagara and Niagara 2 chips, both of which feature an eight-core design. The Niagara 2 supports more threads and operates at 1.6 GHz.

High-end Intel Xeon processors that are on the LGA 771, LGA1336, and LGA 2011 sockets and high-end AMD Opteron processors that are on the C32 and G34 sockets are DP (dual processor) capable, as well as the older Intel Core 2 Extreme QX9775 also used in an older Mac Pro by Apple and the Intel Skulltrail motherboard. AMD's G34 motherboards can support up to four CPUs and Intel's LGA 1567 motherboards can support up to eight CPUs.

The modern desktop sockets do not support systems with multiple CPUs but very few applications outside of the professional market can make good use of more than four cores and both Intel and AMD currently offer fast quad and six core desktop CPUs so this is generally a moot point anyway. AMD also offers the first and still currently the only eight core desktop CPUs with the FX-8xxx line but anything with more than four cores is generally not very useful in home desktops. As of January 24 of 2012, these FX processors are generally inferior to similarly priced and some times cheaper Intel quad core Sandy Bridge models.

The desktop market has been in a transition towards quad core CPUs since Intel's Core 2 Quads were released and now are quite common although dual core CPUs are still more prevalent. This is largely because of people using older or mobile computers, both of which have a much lower chance of having more than two cores than newer desktops and because of how most computer users are not heavy users. AMD offers CPUs with more cores for a given amount of money than similarly priced Intel CPUs but the AMD cores are somewhat slower than Intel cores so the two will trade blows in different applications depending on how well threaded the programs being used are.

For example, Intel's cheapest Sandy Bridge quad core CPUs often cost almost twice as much as AMD's cheapest Athlon II, Phenom II, and FX quad core CPUs but Intel has dual core CPUs in the same price ranges as AMD's cheaper quad core CPUs. In an application that uses one or two threads the Intel dual cores will out-perform AMD's similarly priced quad core CPUs and if a program supports three or four threads the cheap AMD quad core CPUs will out-perform the similarly priced Intel dual-core CPUs.

Historically AMD and Intel have switched places as the company with the fastest CPU several times. Intel is currently winning on the desktop side of the computer CPU market with their Sandy Bridge and upcoming Ivy Bridge series. In servers AMD's new Opterons seem to have superior performance for their price points. This means that AMD is currently more competitive in low to mid end servers and workstations where more cores and threads are more effectively utilized since professional software often uses more, if not all available threads.

RISC

In the mid-1980s to early-1990s, a crop of new high-performance Reduced Instruction Set Computer (RISC) microprocessors appeared, influenced by discrete RISC-like CPU designs such as the IBM 801 and others. RISC microprocessors were initially used in special-purpose machines and Unix workstations, but then gained wide acceptance in other roles.

In 1986, HP released its first system with a PA-RISC CPU. The first commercial RISC microprocessor design was released either by MIPS Computer Systems, the 32-bit R2000 (the R1000 was not released) or by Acorn computers, the 32-bit ARM2 in 1987. The R3000 made the design truly practical, and the R4000 introduced the world's first commercially available 64-bit RISC microprocessor. Competing projects would result in the IBM POWER and Sun SPARC architectures. Soon every major vendor was releasing a RISC design, including the AT&T CRISP, AMD 29000, Intel i860 and Intel i960, Motorola 88000, DEC Alpha.

As of 2007, two 64-bit RISC architectures are still produced in volume for non-embedded applications: SPARC and Power ISA.

Special-purpose designs

A microprocessor is a general purpose system. Several specialized processing devices have followed from the technology. Microcontrollers integrate a microprocessor with periphal devices for control of embedded system. A digital signal processor (DSP) is specialized for signal processing. Graphics processing units may have no, limited, or general programming facilities. For example, GPUs through the 1990s were mostly non-programmable and have only recently gained limited facilities like programmable vertex shaders.

Market statistics

In 2003, about $44 billion (USD) worth of microprocessors were manufactured and sold.[37] Although about half of that money was spent on CPUs used in desktop or laptop personal computers, those count for only about 2% of all CPUs sold.[38]

About 55% of all CPUs sold in the world are 8-bit microcontrollers, over two billion of which were sold in 1997.[39]

As of 2002, less than 10% of all the CPUs sold in the world are 32-bit or more. Of all the 32-bit CPUs sold, about 2% are used in desktop or laptop personal computers. Most microprocessors are used in embedded control applications such as household appliances, automobiles, and computer peripherals. Taken as a whole, the average price for a microprocessor, microcontroller, or DSP is just over $6.[38]

About ten billion CPUs were manufactured in 2008. About 98% of new CPUs produced each year are embedded.[40]

See also

- Arithmetic logic unit
- Central processing unit
- Comparison of CPU architectures
- Computer architecture
- Computer engineering
- CPU design
- Floating point unit
- Instruction set
- List of instruction sets
- List of microprocessors
- Microarchitecture

- Microcode
- Microprocessor chronology

Notes

[1] Osborne, Adam (1980). *An Introduction to Microcomputers*. **Volume 1: Basic Concepts** (2nd ed.). Berkely, California: Osborne-McGraw Hill. ISBN 0-931988-34-9.

[2] Krishna Kant *Microprocessors And Microcontrollers: Architecture Programming And System Design*PHI Learning Pvt. Ltd., 2007 ISBN 8120331915 page 61, describing the iAPX 432

[3] Back to the Moon: The Verification of a Small Microprocessor's Logic Design (http://klabs.org/richcontent/verification/80k85_verification/index.htm) - NASA Office of Logic Design

[4] Moore, Gordon (19 April 1965). "Cramming more components onto integrated circuits" (ftp://download.intel.com/museum/Moores_Law/Articles-Press_Releases/Gordon_Moore_1965_Article.pdf) (PDF). *Electronics* **38** (8). . Retrieved 2009-12-23.

[5] (PDF) *Excerpts from A Conversation with Gordon Moore: Moore's Law* (ftp://download.intel.com/museum/Moores_Law/Video-Transcripts/Excepts_A_Conversation_with_Gordon_Moore.pdf). Intel. 2005. . Retrieved 2009-12-23.

[6] Mack, Pamela E. (30 November 2005). "The Microcomputer Revolution" (http://www.clemson.edu/caah/history/FacultyPages/PamMack/lec122/micro.htm). . Retrieved 2009-12-23.

[7] (PDF) *History in the Computing Curriculum* (http://www.hofstra.edu/pdf/CompHist_9812tla6.PDF). . Retrieved 2009-12-23.

[8] Halfhill, Tom, "The Intel 4004's 35th Anniversary", Microprocessor Report, Dec. 18, 2006, www.MPRonline.com

[9] Faggin, Federico; Hoff, Marcian E., Jr.; Mazor, Stanley; Shima, Masatoshi (December 1996). "The History of the 4004" (http://ieeexplore.ieee.org/xpl/freeabs_all.jsp?arnumber=546561). *IEEE Micro* **16** (6): 10–20. .

[10] Faggin, F.; Klein, T.; L. (23 October 1968). "Insulated Gate Field Effect Transistor Integrated Circuits with Silicon Gates" (http://www.intel4004.com/images/iedm_covart.jpg) (JPEG image). International Electronic Devices Meeting. IEEE Electron Devices Group. . Retrieved 2009-12-23.

[11] Augarten, Stan (1983). *The Most Widely Used Computer on a Chip: The TMS 1000* (http://smithsonianchips.si.edu/augarten/p38.htm). New Haven and New York: Ticknor & Fields. ISBN 0-89919-195-9. . Retrieved 2009-12-23.

[12] http://www.google.com/patents?vid=3757306

[13] http://www.google.com/patents?vid=4074351

[14] McGonigal, James (20 September 2006). "Microprocessor History: Foundations in Glenrothes, Scotland" (http://www.spingal.plus.com/micro). *McGonigal personal website* accessdate=2009-12-23. .

[15] Tout, Nigel. "ANITA at its Zenith" (http://anita-calculators.info/html/anita_at_its_zenith.html). *Bell Punch Company and the ANITA calculators*. . Retrieved 2010-07-25.

[16] 16 Bit Microprocessor Handbook by Gerry Kane, Adam Osborne ISBN 0079310435 (0-07-931043-5)

[17] Holt, Ray M.. "World's First Microprocessor Chip Set" (http://web.archive.org/web/20100725060322/http://www.microcomputerhistory.com/index.php). Ray M. Holt website. Archived from the original (http://www.microcomputerhistory.com) on 2010-07-25. . Retrieved 2010-07-25.

[18] Holt, Ray (27 September 2001). *Lecture: Microprocessor Design and Development for the US Navy F14 FighterJet* (http://www.pdl.cmu.edu/SDI/2001/092701.html) (Speech). Room 8220, Wean Hall, Carnegie Mellon University, Pittsburgh, PA, US. . Retrieved 2010-07-25.

[19] Parab, Jivan S.; Shelake, Vinod G.; Kamat, Rajanish K.; Naik, Gourish M. (2007) (PDF). *Exploring C for Microcontrollers: A Hands on Approach* (http://ee.sharif.edu/~sakhtar3/books/Exploring C for Microcontrollers.pdf). Springer. p. 4. ISBN 978-1-4020-6067-0. . Retrieved 2010-07-25.

[20] Hyatt, Gilbert P., "Single chip integrated circuit computer architecture", Patent 4942516 (http://www.google.com/patents/about?id=cNcbAAAAEBAJ), issued July 17, 1990

[21] "The Gilbert Hyatt Patent" (http://www.intel4004.com/hyatt.htm). *intel4004.com*. Federico Faggin. . Retrieved 2009-12-23.

[22] Crouch, Dennis (1 July 2007). "Written Description: CAFC Finds Prima Facie Rejection(Hyatt v. Dudas (Fed. Cir. 2007)" (http://www.patentlyo.com/patent/2007/07/hyatt-v-dudas-f.html). *Patently-O blog*. . Retrieved 2009-12-23.

[23] Basset, Ross (2003). "When is a Microprocessor not a Microprocessor? The Industrial Construction of Semiconductor Innovation" (http://books.google.com/books?id=rsRJTiu1h9MC). In Finn, Bernard. *Exposing Electronics*. Michigan State University Press. p. 121. ISBN 0-87013-658-5. .

[24] "1971 - Microprocessor Integrates CPU Function onto a Single Chip" (http://www.computerhistory.org/semiconductor/timeline/1971-MPU.html). *The Silicon Engine*. Computer History Museum. . Retrieved 2010-07-25.

[25] Shaller, Robert R. (15 April 2004). "Dissertation: Technological Innovation in the Semiconductor Industry: A Case Study of the International Technology Roadmap for Semiconductors" (http://web.archive.org/web/20061219012629/http://home.comcast.net/~gordonepeterson2/schaller_dissertation_2004.pdf) (PDF). George Mason University. Archived from the original (http://home.comcast.net/~gordonepeterson2/schaller_dissertation_2004.pdf) on 2006-12-19. . Retrieved 2010-07-25.

[26] RW (3 March 1995). "Interview with Gordon E. Moore" (http://www-sul.stanford.edu/depts/hasrg/histsci/silicongenesis/moore-ntb.html). *LAIR History of Science and Technology Collections*. Los Altos Hills, California: Stanford University. .

[27] Bassett 2003. pp. 115, 122.

[28] Ceruzzi, Paul E. (May 2003). *A History of Modern Computing* (2nd ed.). MIT Press. pp. 220–221. ISBN 0262532034.

[29] Wood, Lamont (August 2008). "Forgotten history: the true origins of the PC" (http://www.webcitation.org/5vZKtCs1r). Computerworld. Archived from the original (http://www.computerworld.com/s/article/9111341/Forgotten_PC_history_The_true_origins_of_the_personal_computer) on 2011-01-07. . Retrieved 2011-01-07.

[30] Intel 8087 datasheet, pg. 1

[31] Intel387 DX MATH COPROCESSOR datasheet, pg. 1, March 1992 (Order Number: 240448-005)

[32] "Essentially, the 80C187 can be treated as an additional resource or an extension to the CPU. The 80C186 CPU together with an 80C187 can be used as a single unified system." Intel 80C187 datasheet, p. 3, November 1992 (Order Number: 270640-004).

[33] Priorartdatabase.com (http://priorartdatabase.com/IPCOM/000059679#)

[34] "Shoji, M. Bibliography" (http://cm.bell-labs.com/cm/cs/bib/shoji.bib). Bell Laboratories. 7 October 1998. . Retrieved 2009-12-23.

[35] "Timeline: 1982–1984" (http://www.bell-labs.com/org/physicalsciences/timeline/span23.html). *Physical Sciences & Communications at Bell Labs*. Bell Labs, Alcatel-Lucent. 17 January 2001. . Retrieved 2009-12-23.

[36] Turley, Jim (July 1998). "MCore: Does Motorola Need Another Processor Family?" (http://web.archive.org/web/19980702003323/http://www.embedded.com/98/9807sr.htm). *Embedded Systems Design*. TechInsights (United Business Media). Archived from the original (http://www.embedded.com/98/9807sr.htm) on 1998-07-02. . Retrieved 2009-12-23.

[37] WSTS Board Of Directors. "WSTS Semiconductor Market Forecast World Release Date: 1 June 2004 - 6:00 UTC" (http://web.archive.org/web/20041207091926/http://www.wsts.org/press.html) (Press release). World Semiconductor Trade Statistics. Archived from the original (http://www.wsts.org/press.html) on 2004-12-07. .

[38] Turley, Jim (18 December 2002). "The Two Percent Solution" (http://www.embedded.com/shared/printableArticle.jhtml?articleID=9900861). *Embedded Systems Design*. TechInsights (United Business Media). . Retrieved 2009-12-23.

[39] Cantrell, Tom (1998). "Microchip on the March" (http://web.archive.org/web/20070220134759/http://www.circuitcellar.com/library/designforum/silicon_update/3/index.asp). Archived from the original (http://www.circuitcellar.com/library/designforum/silicon_update/3/) on 2007-02-20. .

[40] Barr, Michael (1 August 2009). "Real men program in C" (http://www.embedded.com/columns/barrcode/218600142?pgno=2). *Embedded Systems Design*. TechInsights (United Business Media). p. 2. . Retrieved 2009-12-23.

References

- Ray, A. K.; Bhurchand, K.M.. *Advanced Microprocessors and Peripherals*. India: Tata McGraw-Hill.

External links

- Dirk Oppelt. "The CPU Collection" (http://www.cpu-collection.de/). Retrieved 2009-12-23.
- Gennadiy Shvets. "CPU-World" (http://www.cpu-world.com/). Retrieved 2009-12-23.
- Jérôme Cremet. "The Cecko's CPU Library" (http://gecko54000.free.fr/). Retrieved 2009-12-23.
- "How Microprocessors Work" (http://computer.howstuffworks.com/microprocessor.htm). Retrieved 2009-12-23.
- William Blair. "IC Die Photography" (http://diephotos.blogspot.com/). Retrieved 2009-12-23.
- John Bayko (December 2003). "Great Microprocessors of the Past and Present" (http://jbayko.sasktelwebsite.net/cpu.html). Retrieved 2009-12-23.
- Wade Warner (22 December 2004). "Great moments in microprocessor history" (http://www-106.ibm.com/developerworks/library/pa-microhist.html?ca=dgr-mw08MicroHistory). IBM. Retrieved 2009-12-23.
- Ray M. Holt. "theDocuments" (http://firstmicroprocessor.com/?page_id=17). *World's First Microprocessor*. Retrieved 2009-12-23.

United_States_Department_of_Defense

Department of Defense	
Department overview	
Formed	August 10, 1949
Preceding Department	Department of War and Department of the Navy
Jurisdiction	Federal government of the United States
Headquarters	The Pentagon, Arlington Co., Virginia 38°52′15.56″N 77°3′21.46″W
Employees	718,000 civilian 1,418,542 Active Duty military, 1,100,000+ reserve: 3.23 million total [1] (2009)
Annual budget	US$530.1 billion (2010)[2] US$549.1 billion (2011)[3] US$553.0 billion (est. 2012)
Department executives	Leon Panetta, Secretary Ashton Carter, Deputy Secretary
Child agencies	Department of the Army Department of the Navy Department of the Air Force
Website	
defense.gov [4]	

The **Department of Defense** (also known as the **Defense Department**, **USDOD**, **DOD**, **DoD** or **the Pentagon**[5]) is the Executive Department of the Government of the United States of America charged with coordinating and supervising all agencies and functions of the government concerned directly with national security and the United States armed forces. The Department is also the largest employer in the world[6] , with more than 2.13 million active duty Soldiers, Sailors, Marines, Airmen, and also civilian workers, and

The Pentagon, headquarters of the United States Department of Defense.

over 1.1 million National Guardsmen and members of the Army, Navy, Air Force, and Marine Reserves. The grand total is just over 3.2 million servicemen and servicewomen, plus the civilians who support them.

The Department – headed by the Secretary of Defense – has three subordinate military departments: the Department of the Army, the Department of the Navy, and the Department of the Air Force. In addition, there are many Defense Agencies such as the Defense Advanced Research Projects Agency (DARPA), the Defense Intelligence Agency (DIA), the Defense Logistics Agency (DLA), the Missile Defense Agency, the National Geospatial-Intelligence Agency (NGA), the National Reconnaissance Office (NRO), the National Security Agency (NSA), and the Pentagon Force Protection Agency (PFPA), all of which are under the command of the Secretary of Defense. DoD also operates several joint services schools, including the National Defense University (NDU) and the National War College (NWC).

The Department is allocated the highest level of budgetary resources among all Federal agencies, and this amounts to more than one-half of the annual Federal discretionary budget.

History

President Harry Truman signs the National Security Act Amendment of 1949

The United States Congress created the War Department in 1789 and the Navy Department in 1798. The secretaries of each of these departments reported directly to the President as cabinet-level advisors.[7]

In a special message to Congress on 19 December 1945, President Harry Truman proposed creation of a unified department of state defense, citing both wasteful military spending and inter-departmental conflicts. Deliberations in Congress went on for months focusing heavily on the role of the military in society and the threat of granting too much military power to the executive.[8]

On July 26, 1947, Truman signed the National Security Act of 1947, which set up a unified military command known as the "National Military Establishment", as well as creating the Central Intelligence Agency, the National Security Council, National Security Resources Board, United States Air Force (formerly the Army Air Forces) and the Joint Chiefs of Staff. The act placed the National Military Establishment under the control of a single Secretary of Defense.[7] [9] [10] The National Military Establishment formally began operations on September 18, the day after the Senate confirmed James V. Forrestal as the first Secretary of Defense.[9] The National Military Establishment was renamed the "Department of Defense" on August 10, 1949, in an amendment to the original 1947 law.[11]

Organizational structure

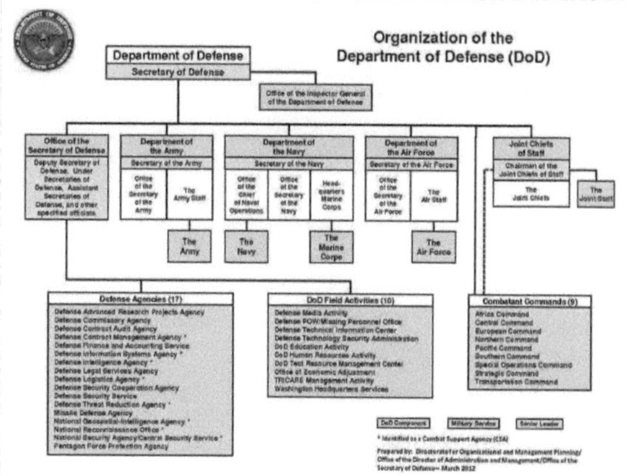

The Secretary of Defense, appointed by the President with the advice and consent of the Senate, is by federal law (10 U.S.C. § 113 [12]) the head of the Department of Defense, "the principal assistant to the President in all matters relating to Department of Defense", and has "authority, direction and control over the Department of Defense". Because the Constitution vests all military authority in Congress and the President, the statutory authority of the Secretary of Defense is derived from their constitutional authorities. Since it is impractical for either Congress or the President to participate in every piece of Department of Defense affairs, the Secretary of Defense, and the Secretary's subordinate officials generally exercise military authority.

The Department of Defense is composed of the Office of the Secretary of Defense (OSD), the Joint Chiefs of Staff (JCS) and the Joint Staff (JS), Office of the Inspector General (DODIG), the Combatant Commands, the Military Departments (Department of the Army (DA), Department of the Navy (DON) & Department of the Air Force (DAF)), the Defense Agencies and DoD Field Activities, the National Guard Bureau (NGB), and such other offices, agencies, activities, organizations, and commands established or designated by law, or by the President or by the Secretary of Defense.

Department of Defense Directive 5100.01 [13] describes the organizational relationships within the Department, and is the foundational issuance for delineating the major functions of the Department. The latest version, signed by former Secretary of Defense Robert Gates in December of 2010, is the first major re-write since 1987.[14]

Office of the Secretary of Defense

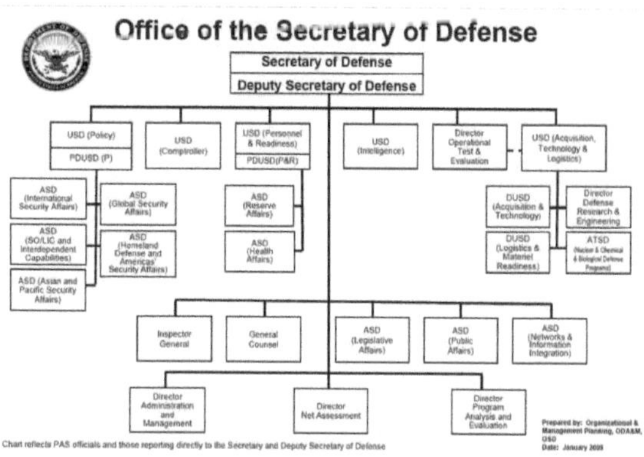

2008 OSD organizational chart

The Office of the Secretary of Defense (OSD) is the Secretary and Deputy Secretary's (mainly) civilian staff.

OSD is the principal staff element of the Secretary of Defense in the exercise of policy development, planning, resource management, fiscal and program evaluation and oversight, and interface and exchange with other U.S. Government departments and agencies, foreign governments, and international organizations, through formal and informal processes. OSD also performs oversight and management of the Defense Agencies and DoD Field Activities.

OSD also supervises the following Department of Defense agencies:

- Armed Forces Radiobiology Research Institute (AFRRI)
- Defense Advanced Research Projects Agency (DARPA)
- Defense Commissary Agency (DeCA)
- Defense Contract Audit Agency (DCAA)
- Defense Contract Management Agency (DCMA)
- Defense Clandestine Service (DCS)
- Defense Finance and Accounting Service (DFAS)
- Defense Information Systems Agency (DISA)
- Defense Intelligence Agency (DIA)
- Defense Legal Services Agency
- Defense Logistics Agency (DLA)
- Defense Security Cooperation Agency (DSCA) (formerly Defense Security Assistance Agency)
- Defense Security Service (DSS) (formerly Defense Investigative Service)
- Defense Technical Information Center (DTIC)
- Defense Threat Reduction Agency (DTRA)
- Missile Defense Agency (MDA)
- National Geospatial-Intelligence Agency (NGA)
- National Reconnaissance Office (NRO)
- National Security Agency (NSA) / Central Security Service (CSS)

Defense Agencies

Defense Agencies
National Security Agency

Defense Intelligence Agency

Defense Logistics Agency

Defense Security Cooperation Agency

National Geospatial-Intelligence Agency

Defense Advanced Research Projects Agency

Joint Chiefs of Staff

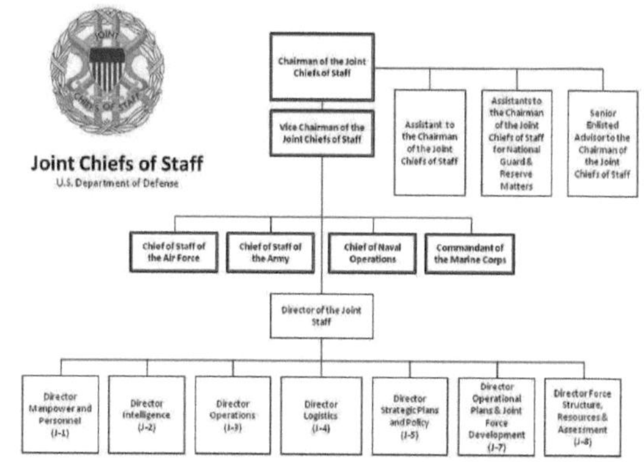

Joint Chiefs of Staff/Joint Staff organizational chart.

The Joint Chiefs of Staff (JCS) is a body of senior uniformed leaders in the Department of Defense who advise the Secretary of Defense, the Homeland Security Council, the National Security Council and the President on military matters. The composition of the Joint Chiefs of Staff is defined by statute and consists of the *Chairman of the Joint Chiefs of Staff* (*CJCS*), *Vice Chairman of the Joint Chiefs of Staff* (*VCJCS*), *Senior Enlisted Advisor to the Chairman* (*SEAC*), and the Military Service Chiefs from the Army, Navy, Air Force and the Marine Corps, all appointed by the President following Senate confirmation.[15] Each of the individual Military Service Chiefs, outside of their Joint Chiefs of Staff obligations, works directly for the Secretary of the Military Department concerned, i.e. Secretary of the Army, Secretary of the Navy, and the Secretary of the Air Force.[16] [17] [18] [19]

Following the Goldwater-Nichols Act in 1986 the Joint Chiefs of Staff do not have operational command authority, neither individually nor collectively, as the chain of command goes from the President to the Secretary of Defense, and from the Secretary of Defense to the Commanders of the Combatant Commands.[20] Goldwater-Nichols also created the office of Vice Chairman, and the Chairman is now designated as the *principal military adviser* to the Secretary of Defense, the Homeland Security Council, the National Security Council and to the President.[21]

The **Joint Staff** (JS) is a headquarters staff in the Pentagon, composed of personnel from all the four services, that assists the Chairman and the Vice Chairman in discharging their responsibilities and is managed by the *Director of the Joint Staff* (*DJS*) who is a Lieutenant General or Vice Admiral.[22] [23]

Military Departments

Military Departments of the DOD

Department of the Army

Department of the Navy

Department of the Air Force

There are three Military Departments within the Department of Defense:

1. the Department of the Army, of which the United States Army is organized within.
2. the Department of the Navy, of which the United States Navy and the United States Marine Corps are organized within.
3. the Department of the Air Force, of which the United States Air Force is organized within.

The Military Departments are each headed by their own Secretary (i.e., Secretary of the Army, Secretary of the Navy and Secretary of the Air Force), appointed by the President, with the advice and consent of the Senate. They have legal authority under Title 10 of the United States Code to conduct all the affairs of their respective departments within which the military services are organized.[24] The Secretaries of the Military Departments are (by law) subordinate to the Secretary of Defense and (by SecDef delegation) to the Deputy Secretary of Defense.

The Secretaries of the Military Departments, in turn, normally exercises authority over their forces by delegation through their respective Service Chiefs (i.e., Chief of Staff of the Army, Chief of Naval Operations, Commandant of the Marine Corps, and Chief of Staff of the Air Force) over forces not assigned to a Combatant Command.[25]

The Secretaries of the Military Departments and the Service Chiefs do not possess operational command authority over U.S. troops (this power was stripped from them in the Defense Reorganization Act of 1958), and instead the Military Departments are tasked solely with "the training, provision of equipment, and administration of troops."[25]

Unified Combatant Commands

A Unified Combatant Command is a single force composed of personnel and equipment from at least two Military Departments, which has a broad and continuing mission.[26] [27]

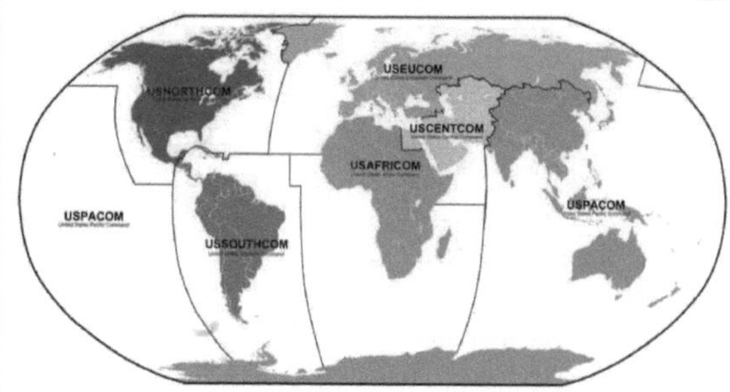

Map of the DoD's geographic commands

The Military Departments are responsible for equipping and training the troops to fight, while the Unified Combatant Commands are responsible for actual operational command of military forces.[27] Almost all operational U.S. forces are under the authority of a Unified Command.[25] The Unified Commands are governed by a Unified Command Plan, a frequently updated document (produced by the DOD) which lays out the Command's mission, geographical/functional responsibilities, and force structure.[27]

During military operations, the chain of command runs from the President to the Secretary of Defense to the combatant commanders of the Combatant Commands.[25]

The United States currently has 9 Combatant Commands, organized either on a geographical basis (known as "Area of Responsibility", AOR) or on a global, functional basis[28] :

- U.S. Northern Command (USNORTHCOM)
- U.S. Southern Command (USSOUTHCOM)
- U.S. Central Command (USCENTCOM)
- U.S. European Command (USEUCOM)
- U.S. Pacific Command (USPACOM)
- U.S. Africa Command (USAFRICOM)
- U.S. Strategic Command (USSTRATCOM)
- U.S. Special Operations Command (USSOCOM)
- U.S. Transportation Command (USTRANSCOM)

Expenditures

Department of Defense spending in 2010 was 4.8% of GDP and accounted for approximately 45% of budgeted global military spending – more than the next 17 largest militaries combined.[29] [30]

The Department of Defense accounts for the majority of federal discretionary spending. In FY 2010 the DOD budgeted spending accounted for 21% of the U.S. Federal Budget, and 53% of federal discretionary spending, which represents funds not accounted for by pre-existing obligations.[31] [32]

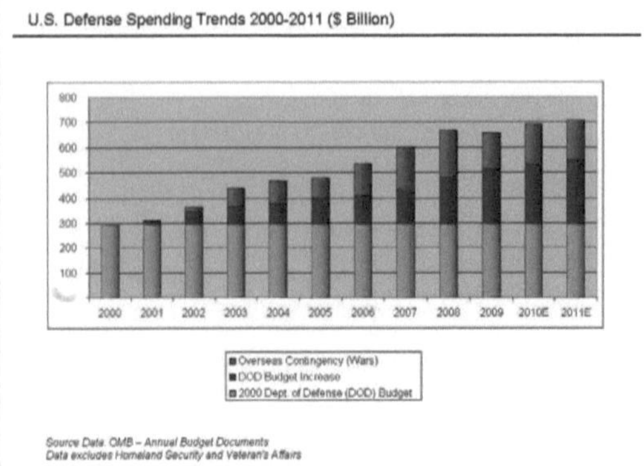

Chart showing growth in U.S. DoD spending 2000–2011

In the 2010 United States federal budget, the DoD was allocated a base budget of $533.7 billion, with a further $75.5 billion adjustment in respect of 2009, and $130 billion for overseas contingencies.[33] The subsequent 2010 DoD Financial Report shows DoD total budgetary resources for fiscal year 2010 were $1.2 trillion.[34] Of these resources, $1.1 trillion were obligated and $994 billion were disbursed, with the remaining resources relating to multi-year modernization projects requiring additional time to procure.[34] After over a decade of non-compliance, Congress has established a deadline of FY 2017 for the DoD to achieve audit readiness.[35]

Energy use

The Department of Defense was the largest single consumer of energy in the United States in 2006.[36]

In FY 2006, the Department used almost 30,000 gigawatt hours (GWH) of electricity, at a cost of almost $2.2 billion. The DoD's electricity use would supply enough electricity to power more than 2.6 million average American homes. In electricity consumption, if it were a country, the DoD would rank 58th in the world, using slightly less than Denmark and slightly more than Syria (CIA World Factbook, 2006).[37]

The DOD is responsible for 93% of all US government fuel consumption in 2007 (Department of the Air Force: 52%; Department of the Navy: 33%; Department of the Army: 7%; other DoD components: 1%).[37] The Department of Defense uses 4600000000 US gallons (**expected operatorexpected operatorexpected operatorexpected operatorexpected operatorexpected operatorexpected operatorexpected operatorexpected operatorexpected operatorexpected operatorexpected operatorexpected operatorexpected operator**$\times 10^{10}$ L) of fuel annually, an average of 12600000 US gallons (**unknown operator: u'strong'** L) of fuel per day. A large Army division may use about 6000 US gallons (**unknown operator: u'strong'** L) per day. According to the 2005 *CIA World Factbook*, if it were a country, the DoD would rank 34th in the world in average daily oil use, coming in just behind Iraq and just ahead of Sweden.[38] The Air Force is the largest user of fuel energy in the federal government. The Air Force uses 10% of the nation's aviation fuel. (JP-8 accounts for nearly 90% of its fuels.) This fuel usage breaks down as such: 82% jet fuel, 16% facility management and 2% ground vehicle/equipment.[39]

Related legislation

The organization and functions of the DoD are set forth in Title 10 of the United States Code.

Other significant legislation related to the Department of Defense includes:

- 1947: National Security Act of 1947
- 1958: Department of Defense Reorganization Act PL 85-899
- 1963: Department of Defense Appropriations Act PL 88-149
- 1963: Military Construction Authorization Act PL 88-174
- 1967: Supplemental Defense Appropriations Act PL 90-8
- 1984: Department of Defense Authorization Act PL 98-525
- 1986: Goldwater-Nichols Act of 1986 or Department of Defense Reorganization Act PL 99-433
- 1996: Antiterrorism and Effective Death Penalty Act PL 104-132

See also

- Defense contractor
- Defense industry
- List of United States military bases
- Military-industrial complex
- Nuclear Weapons
- Private military contractor
- U.S. Department of Homeland Security
- U.S. Department of Justice

References

[1] "Department of Defense" (http://siadapp.dmdc.osd.mil/personnel/MILITARY/history/hst0909.pdf) (PDF). . Retrieved 2010-10-15.
[2] "Budget of the US Government, FY 2011" (http://www.gpoaccess.gov/usbudget/fy12/pdf/BUDGET-2012-BUD-7.pdf) (PDF). . Retrieved 2010-10-15.
[3] Defense.gov (http://www.gpoaccess.gov/usbudget/fy12/pdf/BUDGET-2012-BUD-7.pdf)
[4] http://www.defense.gov/
[5] Initially the National Military Establishment (NME)
[6] "Which is the world's biggest employer?" (http://www.bbc.co.uk/news/magazine-17429786). *BBC News*. March 20, 2012. .
[7] Polmar, Norman (2005). *The Naval Institute guide to the ships and aircraft of the U.S. fleet* (http://books.google.com/books?id=8MwyTX-iA2wC&pg=PA17). Naval Institute Press. p. 17. ISBN 978-1-59114-685-8. .
[8] Hogan, Michael J. (2000). *A cross of iron: Harry S. Truman and the origins of the national security state, 1945-1954* (http://books.google.com/books?id=Ild4C3cY7Y7lC&pg=PA37). Cambridge University Press. pp. 37–38. ISBN 978-0-521-79537-1. .
[9] Hugh W. Nibley (1944-07-04). "Warfare and the Book of Mormon" (http://maxwellinstitute.byu.edu/publications/transcripts/?id=14). Maxwellinstitute.byu.edu. . Retrieved 2010-10-15.
[10] Bolton, M. Kent (2008). *U.S. national security and foreign policymaking after 9/11: present at the re-creation* (http://books.google.com/books?id=IETboc9ajpQC&pg=PA3). Rowman & Littlefield. p. 3. ISBN 978-0-7425-5900-4. .
[11] Rearden, Steven L. (2001). "Department of Defense" (http://books.google.com/books?id=h7dG-pVarDAC&pg=PA439). In DeConde, Alexander et al.. *Encyclopedia of American Foreign Policy, Volume 1*. Simon and Schuster. ISBN 978-0-684-80657-0. .
[12] http://www.law.cornell.edu/uscode/10/113.html
[13] http://www.dtic.mil/whs/directives/corres/pdf/510001p.pdf
[14] (http://odam.defense.gov/omp/Functions/Organizational_Portfolios/Evolution of 5100.1.html)
[15] (http://www.law.cornell.edu/uscode/html/uscode10/usc_sec_10_00000151----000-.html) 10 USC 151. Joint Chiefs of Staff: composition; functions
[16] 10 U.S.C. §3033 (http://uscode.house.gov/download/pls/10C305.txt)
[17] 10 U.S.C. §5033 (http://uscode.house.gov/download/pls/10C505.txt)
[18] 10 U.S.C. §5043 (http://uscode.house.gov/download/pls/10C506.txt)
[19] 10 U.S.C. §8033 (http://uscode.house.gov/download/pls/10C805.txt)
[20] 10 U.S.C. §162(b) (http://uscode.house.gov/download/pls/10C6.txt)
[21] 10 U.S.C §151(b) (http://uscode.house.gov/download/pls/10C5.txt)

[22] 10 U.S.C §155 (http://uscode.house.gov/download/pls/10C5.txt)

[23] Polmar, Norman (2005). "Defense organization" (http://books.google.com/books?id=8MwyTX-iA2wC&pg=PA19). *The Naval Institute guide to the ships and aircraft of the U.S. fleet*. Naval Institute Press. ISBN 978-1-59114-685-8. .

[24] 10 U.S.C. § 3013 (http://www.law.cornell.edu/uscode/10/3013.html), 10 U.S.C. § 5013 (http://www.law.cornell.edu/uscode/10/5013.html) and 10 U.S.C. § 8013 (http://www.law.cornell.edu/uscode/10/8013.html)

[25] Polmar, Norman (2005). "Defense Organization" (http://books.google.com/books?id=8MwyTX-iA2wC&pg=PA20). *The Naval Institute guide to the ships and aircraft of the U.S. fleet*. Naval Institute Press. p. 20. ISBN 978-1-59114-685-8. .

[26] Watson, Cynthia A. (2010). *Combatant Commands: Origins, Structure, and Engagements* (http://books.google.com/books?id=v183dxvXOa0C&pg=PA3). ABC-CLIO. p. 3. ISBN 978-0-313-35432-8. .

[27] Whitley, Joe D. et al., ed. (2009). "Unified Combatant Commands and USNORTHCOM" (http://books.google.com/books?id=bJI54yr1ymQC&pg=PA44). *Homeland security: legal and policy issues*. American Bar Association. ISBN 978-1-60442-462-1. .

[28] Reveron, Derek S. (2007). *America's Viceroys: The Military and U.S. Foreign Policy* (http://books.google.com/books?id=OsnxXXxn5BEC&pg=PA26). Macmillan. p. 26. ISBN 978-0-230-60219-9. .

[29] "Military Spending: Defence Costs" (http://www.economist.com/blogs/dailychart/2011/06/military-spending). *The Economist*. June 8, 2011. .

[30] "The SIPRI Military Expenditure Database" (http://milexdata.sipri.org/). Stockholm International Peace Research Institute. . Retrieved 13 December 2011.

[31] "United States Federal Budget for Fiscal Year 2010" (http://www.gpoaccess.gov/usbudget/fy10/pdf/fy10-newera.pdf). Government Printing Office. . Retrieved 9 January 2010.

[32] This does not include many military-related items that are outside of the Defense Department budget, such as nuclear weapons research, maintenance, cleanup, and production, which is in the Department of Energy budget, Veterans Affairs, the Treasury Department's payments in pensions to military retirees and widows and their families, interest on debt incurred in past wars, or State Department financing of foreign arms sales and militarily-related development assistance. Neither does it include defense spending that is not military in nature, such as the Department of Homeland Security, counter-terrorism spending by the FBI, and intelligence-gathering spending by NASA.

[33] "United States Federal Budget for Fiscal Year 2010 (*vid.* p.53)" (http://www.gpoaccess.gov/usbudget/fy10/pdf/fy10-newera.pdf). Government Printing Office. . Retrieved 9 January 2010.

[34] "FY 2010 DoD Agencywide Agency Financial Report (*vid.* p.25)" (http://comptroller.defense.gov/cfs/fy2010/01_DoD_Agency-Wide/Fiscal_Year_2010_DoD_Agencywide_Agency Financial Report.pdf). US Department of Defense. . Retrieved 7 January 2011.

[35] "Financial Improvement and Audit Readiness (FIAR) Plan Status Report" (http://comptroller.defense.gov/FIAR/documents/FIAR_Plan_May_2011.pdf). Comptroller, Department of Defense. . Retrieved 13 December 2011.

[36] Andrews Anthony (2011). *Department of Defense Facilities: Energy Conservation Policies and Spending* (http://books.google.com/books?id=s2xbNsaxD4IC&pg=PA1). DIANE Publishing. p. 1. ISBN 978-1-4379-3835-7. .

[37] Colonel Gregory J. Lengyel, USAF, The Brookings Institution, Department of Defense Energy Strategy, August 2007.

[38] Colonel Gregory J. Lengyel, USAF, The Brookings Institution, Department of Defense Energy Strategy, August 2007, (http://www.brookings.edu/~/media/Files/rc/papers/2007/08defense_lengyel/lengyel20070815.pdf)

[39] Powering America's Defense: Energy and the Risks to National Security (http://www.cna.org/documents/PoweringAmericasDefense.pdf), CNA Analysis & Solutions, May 2009

External links

- United States Department of Defense Official website (http://www.defense.gov/)
- Office of the Under Secretary of Defense (Comptroller) (http://comptroller.defense.gov/) Budget and Financial Management Policy
- Read Congressional Research Service (CRS) Reports regarding the Department of Defense (http://digital.library.unt.edu/govdocs/crs/search.tkl?q="department+of+defense"&search_crit=subject&search=Search&date1=Anytime&date2=Anytime&type=form)
- Entire Collection of DoD Freedom of Information Act (FOIA) Reading Room (http://community.theblackvault.com/articles/category/DoD-Archive)
- Budget info (http://www.defenselink.mil/comptroller/defbudget/fy2006/)
- Death and Taxes: 2009 (http://www.wallstats.com/deathandtaxes/resource/) A visual guide and infographic of the 2009 United States federal budget including the Department of Defense with data provided by the Comptrollers office.
- DoD YouTube channel (http://uk.youtube.com/user/DODvClips)
- Proposed and finalized federal regulations from the United States Department of Defense (http://openregs.com/agencies/view/8/department_of_defense)

- DoD IA Policy Chart (http://iac.dtic.mil/iatac/ia_policychart.html)
- U.S. Reporting Agency (http://www.usreportingagency.org)
- U.S. Reporting Agency Groundsquad Homepage (http://www.angelgwinfoundation.com/usreportingagency/groundsquad/index.htm)
- Defense Logistics Agency (http://www.dla.mil/)
- United States Department of Defense (http://www.whorunsgov.com/institutions/defense/) at *WhoRunsGov* at *The Washington Post*

[[zh:]

General_Electric

Type	Public
Traded as	NYSE: GE [1] NASDAQ: GE [2] Dow Jones Component S&P 500 Component Dow Jones Composite Component The Global Dow Component
Industry	Conglomerate
Founded	Schenectady, New York, U.S. (1892)
Founder(s)	Charles Coffin, Edwin Houston, Elihu Thomson, Thomas Edison
Headquarters	Fairfield, Connecticut, U.S.[3]
Area served	Worldwide
Key people	Jeffrey Immelt (Chairman & CEO)
Products	Appliances, aviation, consumer electronics, electrical distribution, electric motors, energy, entertainment, finance, gas, healthcare, lighting, locomotives, oil, software, water, weapons, wind turbines
Revenue	▼ US$ 147.300 billion (2011)[4]
Operating income	▲ US$ 19.155 billion (2011)[4]
Net income	▲ US$ 14.155 billion (2011)[4]
Total assets	▼ US$ 717.242 billion (2011)[4]
Total equity	▼ US$ 118.134 billion (2011)[4]
Employees	301,000 (2011)[4]
Subsidiaries	GE Capital, GE Energy, GE Home & Business Solutions, GE Technology Infrastructure, GE Aviation NBCUniversal (49%)[5]
Website	GE.com [6]

Street entrance to GE Headquarters in Fairfield, Connecticut

General Electric Company (NYSE: GE [1]), or **GE**, is an American multinational conglomerate corporation incorporated in Schenectady, New York and headquartered in Fairfield, Connecticut, United States.[3] [7] The company operates through four segments: Energy, Technology Infrastructure, Capital Finance and Consumer & Industrial.[8] [9]

In 2011, *Fortune* ranked GE the 6th largest firm in the U.S.,[10] as well as the 14th most profitable.[11] Other rankings for 2011/2012 include No. 7 company for leaders (*Fortune*), No. 5 best global brand (*Interbrand*), No. 63 green company (*Newsweek*), No. 15 most admired company (*Fortune*), and No. 19 most innovative company (*Fast Company*).[12]

History

Further information: General Electric timeline

Formation

By 1890, Thomas Alva Edison had brought together several of his business interests under one corporation to form Edison General Electric. At about the same time, Thomson-Houston Electric Company, under the leadership of Charles Coffin, gained access to a number of key patents through the acquisition of a number of competitors. Subsequently, General Electric was formed by the 1892 merger of Edison General Electric of Schenectady, New York and Thomson-Houston Electric Company of Lynn, Massachusetts and both plants remain in operation under the GE banner to this day.[13] The company was incorporated in New York, with the Schenectady plant as headquarters for many years thereafter. Around the same time, General Electric's Canadian counterpart, Canadian General Electric, was formed.

Public company

In 1896, General Electric was one of the original 12 companies listed on the newly formed Dow Jones Industrial Average and still remains after 115 years, the only one remaining on the Dow index. General Electric has however not been on the DOW continuously.[14]

23 Ton diesel electric locomotive made at the General Electric Corp. plant in Schenectady, New York

In 1911 the National Electric Lamp Association (NELA) was absorbed into General Electric's existing lighting business. GE then established its lighting division headquarters at Nela Park in East Cleveland, Ohio. Nela Park is still the headquarters for GE's lighting business.

RCA

The Radio Corporation of America (RCA) was founded by GE in 1919 to further international radio. GE used RCA as its retail arm for radio sales from 1919, when GE began production, until separation in 1930.[15] RCA would quickly grow into an industrial giant of its own.

Power generation

GE's long history of working with turbines in the power-generation field gave them the engineering know-how to move into the new field of aircraft turbosuperchargers. Led by Sanford Alexander Moss, GE introduced the first superchargers during World War I, and continued to develop them during the Interwar period. Superchargers became indispensable in the years immediately prior to World War II, and GE was the world leader in exhaust-driven supercharging when the war started. This experience, in turn, made GE a natural selection to develop the Whittle W.1 jet engine that was demonstrated in the United States in 1941. Although their early work with Whittle's designs was later handed to Allison Engine Company, GE Aviation emerged as one of the world's largest engine manufacturers, second only to the well-founded and older British company, Rolls-Royce plc, which led the way in the design and manufacture of innovative, reliable, efficient, high-performance, heavy-duty jet engines.

In 2002, GE acquired the windpower assets of Enron during its bankruptcy proceedings.[16] Enron Wind was the only surviving U.S. manufacturer of large wind turbines at the time, and GE increased engineering and supplies for the Wind Division and doubled the annual sales to $1.2 billion in 2003.[17] It acquired ScanWind in 2009.[18]

Some consumers boycotted GE light bulbs, refrigerators and other products in the 1980s and 1990s to protest GE's role in nuclear weapons production.[19]

Computing

GE was one of the eight major computer companies during the 1960s — with IBM, the largest, called "Snow White" followed by the "Seven Dwarfs": Burroughs, NCR, Control Data Corporation, Honeywell, RCA, UNIVAC and GE.

GE had an extensive line of general purpose and special purpose computers. Among them were the GE 200, GE 400, and GE 600 series general purpose computers, the GE 4010, GE 4020, and GE 4060 real time process control computers, the Datanet 30 and Datanet 355 message switching computers (Datanet 30 and 355 were also used as front end processors for GE mainframe computers). A Datanet 500 computer was designed, but never sold.

In 1962, GE started developing its GECOS (later renamed GCOS) operating system, originally for batch processing, but later extended to timesharing and transaction processing. Versions of GCOS are still in use today.

In 1964–1969, GE and Bell Laboratories (which soon dropped out) joined with MIT to develop the pioneering and influential Multics operating system on the GE 645 mainframe computer. The project took longer than expected and was not a major commercial success, but it demonstrated important concepts such as single level store, dynamic linking, hierarchical file system, and ring-oriented security. Active development of Multics continued until 1985.

It has been said that GE got into computer manufacturing because in the 1950s they were the largest user of computers outside of the United States federal government. However, in 1970, GE sold its computer division to Honeywell, exiting the computer manufacturing industry, though it retained its timesharing operations for some years afterwards. GE was a major provider of computer timesharing services, through General Electric Information Services (GEIS, now GXS), offering online computing services that included GEnie.

Acquisitions

For a complete list of acquisitions and divestitures, see General Electric timeline

In 1986 GE reacquired RCA, primarily for the NBC television network (also parent of Telemundo Communications Group). The remainder was sold to various companies, including Bertelsmann (Bertelsmann acquired RCA Records) and Thomson SA which traces its roots to Thomson-Houston, one of the original components of GE.

Also in 1986, Kidder, Peabody & Co. a U.S.-based securities firm was sold to GE and following heavy losses was subsequently sold to PaineWebber in 1994.[20]

In 2002 Francisco Partners and Norwest Venture Partners acquired a division of GE called GE Information Systems (GEIS). The new company, named GXS, is based in Gaithersburg, Maryland. GXS is a leading provider of B2B e-Commerce solutions. GE maintains a minority ownership position in GXS.

Also in 2002, GE bought wind turbine manufacturing assets of Enron Wind after the Enron scandals.[21]

In 2004 GE bought 80% of Universal Pictures from Vivendi. Vivendi bought 20% of NBC forming the company NBCUniversal. GE then owned 80% of NBC Universal and Vivendi owned 20%. As of January 28, 2011 GE owns 49% and Comcast 51%.

In 2004 GE completed the spin-off of most of its mortgage and life insurance assets into an independent company, Genworth Financial, based in Richmond, Virginia.

Genpact formerly known as GE Capital International Services (GECIS) was established by GE in late 1997 as its captive India based BPO. GE sold 60% stake in Genpact to General Atlantic and Oak Hill Capital Partners in 2005 and hived off Genpact into an independent business. GE is still a major client to Genpact getting its services in customer service, finance, information technology and analytics.

GE Plastics was sold in 2007 to SABIC.

In May 2007, GE acquired Smiths Aerospace for $4.8 billion.

In May 2008, GE announced it was exploring options for divesting the bulk of its Consumer and Industrial business.

General Electric's Schenectady, New York facilities (including GE's original headquarters) are assigned the ZIP code 12345. (All Schenectady ZIP codes begin with 123, but no others begin with 1234.)

On December 3, 2009, it was announced that NBCUniversal will become a joint venture between GE and cable television operator Comcast. The cable giant will hold a controlling interest in the company, while GE retains a 49% stake and will buy out shares currently owned by Vivendi.[22]

Vivendi will sell its 20% stake in NBCUniversal to GE for US$5.8 billion. Vivendi will sell 7.66% of NBCUniversal to GE for US$2 billion if the GE/Comcast deal is not completed by September 2010 and then sell the remaining 12.34% stake of NBCUniversal to GE for US$3.8 billion when the deal is completed or to the public via an IPO if the deal is not completed.[23] [24]

On March 1, 2010, General Electric (GE) announced that the company is planning to sell its 20.85% stake in Turkey-based Garanti Bank.[25]

In August 2010, GE Healthcare signed a strategic partnership to bring cardiovascular Computed Tomography (CT) technology from start-up Arineta Ltd. of Israel to the hospital market.[26]

In October 2010, General Electric acquired gas engines manufacture Dresser Inc. for a $3 billion deal and also bought a $1.6 billion portfolio of retail credit cards from Citigroup Inc. This is the first major deal since the start of the financial crisis.[27]

On October 14, 2010, GE announced acquisition of data migration & SCADA simulation specialists Opal Software.[28]

December 2010: For the second times of this year (after Dresser acquisition), General Electric Co. buy oil sector company British Wellstream Holding Plc. an oil drilling pipe maker for 800 million pounds ($1.3 billion).[29]

February 2011: The company has agreed to buy the well-support division of John Wood Group Plc for about $2.8 billion. It is another aggressive moves recently of GE Oil & Gas made GE's acquisition was the largest of oil-service unit world wide in 2010.[30]

March 2011: GE announced it has completed the acquisition of privately held Lineage Power Holdings, Inc., from The Gores Group, LLC. [31]

GE Capital sold its $2 billion dollar Mexican assets to Santander for $162 million and exit the business in Mexico. Santander will additionally assume the portfolio debts of GE Capital in the country. The transaction will be finished at first half of 2011. GE Capital will focus in the core business and will shed its non-core assets.[32]

Corporate affairs

Classic GE neon sign

GE is a multinational conglomerate headquartered in Fairfield, Connecticut. Its New York main offices are located at 30 Rockefeller Plaza in Rockefeller Center, known as the GE Building for the prominent GE logo on the roof. NBC's headquarters and main studios are also located in the building. Through its RCA subsidiary, it has been associated with the Center since its construction in the 1930s.

The company describes itself as composed of a number of primary business units or "businesses." Each unit is itself a vast enterprise, many of which would, even as a standalone company, rank in the Fortune 500. The list of GE businesses varies over time as the result of acquisitions, divestitures and reorganizations. GE's tax return is the largest return filed in the United States; the 2005 return was approximately 24,000 pages when printed out, and 237 megabytes when submitted electronically.[33] The company also "spends more on U.S. lobbying than any other company."[34]

In 2005 GE launched its "*Ecomagination*" initiative in an attempt to position itself as a "green" company. GE is currently one of the biggest players in the wind power industry, and it is also developing new environment-friendly products such as hybrid locomotives, desalination and water reuse solutions, and photovoltaic cells. The company "plans to build the largest solar-panel-making factory in the U.S.,"[34] and has set goals for its subsidiaries to lower their greenhouse gas emissions.[35]

On May 21, 2007, GE announced it would sell its GE Plastics division to petrochemicals manufacturer SABIC for net proceeds of $11.6 billion. The transaction took place on August 31, 2007, and the company name changed to SABIC Innovative Plastics, with Brian Gladden as CEO.[36]

CEO

Jeffrey Immelt is the current chairman of the board and chief executive officer of GE. He was selected by GE's Board of Directors in 2000 to replace John Francis Welch Jr. (Jack Welch) following his retirement. Previously, Immelt had headed GE's Medical Systems division (now GE Healthcare) as its President and CEO.

His tenure as the Chairman and CEO started at a time of crisis — he took over the role on September 7, 2001[37] four days before the terrorist attacks on the United States, which killed two employees and cost GE's insurance business $600 million — as well as having a direct effect on the company's Aircraft Engines sector. Immelt has also been selected as one of President Obama's financial advisors concerning the economic rescue plan.

Businesses

GE's divisions include GE Capital, GE Energy, GE Technology Infrastructure, and GE Home & Business Solutions

Through these businesses, GE participates in a wide variety of markets including the generation, transmission and distribution of electricity (e.g. nuclear, gas and solar), lighting, industrial automation, medical imaging equipment, motors, railway locomotives, aircraft jet engines, and aviation services. It co-owns NBCUniversal with Comcast. Through GE Commercial Finance, GE Consumer Finance, GE Equipment Services, and GE Insurance it offers a range of financial services as well. It has a presence in over 100 countries.

GE also produces General Imaging digital cameras.[39] In 2010, General Imaging released the Bridge Camera GE X5 with 14MP and 15x optical zoom.[40] In 2011, it is replaced by 16MP GE X500 with optional red color in Japan besides traditional black or white color in world wide.[41]

GE gauges to control a railway locomotive at a museum near Saskatoon, Canada[38]

Since over half of GE's revenue is derived from financial services, it is arguably a financial company with a manufacturing arm. It is also one of the largest lenders in countries other than the United States, such as Japan. Even though the first wave of conglomerates (such as ITT Corporation, Ling-Temco-Vought, Tenneco, etc.) fell by the wayside by the mid-1980s, in the late 1990s, another wave (consisting of Westinghouse, Tyco, and others) tried and failed to emulate GE's success.

It was announced on May 4, 2008 that GE would auction off its appliances business for an expected sale of $5–8 billion.[42] However, this plan fell through as a result of the recession.

Corporate recognition and rankings

In 2011, *Fortune* ranked GE the 6th largest firm in the U.S.,[10] as well as the 14th most profitable.[11] Other rankings for 2011/2012 include the following:[12]

- #7 company for leaders (*Fortune*)
- #5 best global brand (*Interbrand*)
- #82 green company (*Newsweek*)
- #15 most admired company (*Fortune*)
- #19 most innovative company (*Fast Company*).

For 2010, GE's brand was valued at $42.8 billion.[43] CEO Jeffrey Immelt had a set of changes in the presentation of the brand commissioned in 2004, after he took the reins as chairman, to unify the diversified businesses of GE. The changes included a new corporate color palette, small modifications to the GE logo, a new customized font (GE Inspira), and a new slogan, *"imagination at work"* replacing the longtime slogan *"we bring good things to life"*, composed by David Lucas. The standard requires many headlines to be lowercased and adds visual "white space" to documents and advertising to promote an open and approachable company. The changes were designed by Wolff Olins and are used extensively on GE's marketing, literature and website.

Recent controversies

The six reactors in the 2011 Fukushima I Nuclear Power Plant catastrophe had been designed by General Electric, the design of which had been criticised as far back as 1972.[44]

In March 2011, *The New York Times* reported that, despite earning $14.2 billion in worldwide profits, including more than $5 billion from U.S. operations, General Electric did not owe taxes in 2010. General Electric had a tax benefit of $3.2 billion. This same article also pointed out that GE has reduced its American workforce by one fifth since 2002.[45]

In December 2011, the non-partisan organization Public Campaign criticized General Electric for spending $84.35 million on lobbying and not paying any taxes during 2008–2010, instead getting $4.7 billion in tax rebates, despite making a profit of $10.4 billion, laying off 4,168 workers since 2008, and increasing executive pay by 27% to $75.9 million in 2010 for the top 5 executives.[46]

Environmental record

Pollution

GE has a history of some of its activities giving rise to large-scale air and water pollution. Based on year 2000 data,[47] researchers at the Political Economy Research Institute listed the corporation as the fourth-largest corporate producer of air pollution in the United States, with more than 4.4 million pounds per year (2,000 tons) of toxic chemicals released into the air.[48] GE has also been implicated in the creation of toxic waste. According to EPA documents, only the United States Government, Honeywell, and Chevron Corporation are responsible for producing more Superfund toxic waste sites.[49]

In 1983, New York State Attorney General Robert Abrams filed suit in the United States District Court for the Northern District of New York to compel GE to pay for the cleanup of what was claimed to be more than 100,000 tons of chemicals dumped (legally, at the time) from their plant in Waterford, New York.[50] In 1999, the company agreed to pay a $250 million settlement in connection with claims it polluted the Housatonic River (Pittsfield, Massachusetts) and other sites with polychlorinated biphenyls (PCBs) and other hazardous substances.[51]

From approximately 1947 to 1977, GE discharged as much as 1.3 million pounds of PCBs from its capacitor manufacturing plants at the Hudson Falls and Fort Edward upstate New York facilities into the Hudson River.[52] GE fought a media and political battle to avoid cleaning up the river and countered that dredging the river would actually stir up PCBs.[53] In 2002, GE was ordered to clean up a 40-mile (**unknown operator: u'strong'** km) stretch of the Hudson River it had contaminated.[54]

In 2003, acting on concerns that the plan proposed by GE did not "provide for adequate protection of public health and the environment," the United States Environmental Protection Agency issued a unilateral administrative order for the company to "address cleanup at the GE site" in Rome, Georgia, also contaminated with PCBs.[55]

The nuclear reactors involved in the 2011 crisis at Fukushima I in Japan are GE designs,[56] and the architectural designs were done by Ebasco,[57] formerly owned by GE. Concerns over the design and safety of these reactors were raised as early as 1975.[58]

Environmental initiatives

On June 6, 2011, GE announced that it has licensed solar thermal technology from California-based eSolar for use in power plants that use both solar and natural gas.[59]

On May 26, 2011, GE unveiled its EV Solar Carport, a carport that incorporates solar panels on its roof, with electric vehicle charging stations under its cover.[60]

In May 2005 GE announced the launch of a program called "Ecomagination," intended, in the words of CEO Jeff Immelt "to develop tomorrow's solutions such as solar energy, hybrid locomotives, fuel cells, lower-emission aircraft engines, lighter and stronger durable materials, efficient lighting, and water purification technology".[61] The announcement prompted an op-ed piece in *The New York Times* to observe that, "while General Electric's increased emphasis on clean technology will probably result in improved products and benefit its bottom line, Mr. Immelt's credibility as a spokesman on national environmental policy is fatally flawed because of his company's intransigence in cleaning up its own toxic legacy."[62]

GE has said that it will invest $1.4 billion in clean technology research and development in 2008 as part of its Ecomagination initiative. As of October 2008, the scheme had resulted in 70 green products being brought to market, ranging from halogen lamps to biogas engines. In 2007, GE raised the annual revenue target for its Ecomagination initiative from $20 billion in 2010 to $25 billion following positive market response to its new product lines.[63] In 2010, GE continued to raise its investment by adding $10 billion into Ecomagination over the next five years.[64]

GE (General Electric) Energy's renewable energy business has expanded greatly, to keep up with growing U.S. and global demand for clean energy. Since entering the renewable energy industry in 2002, GE has invested more than

$850 million in renewable energy technology. In August 2008 it acquired Kelman Ltd,[65] a Northern Ireland company specializing in advanced monitoring and diagnostics technologies for transformers used in renewable energy generation, and announced an expansion of its business in Northern Ireland in May 2010.[66] In 2009, GE's renewable energy initiatives, which include solar power, wind power and GE Jenbacher gas engines using renewable and non-renewable methane-based gases,[67] employ more than 4,900 people globally and have created more than 10,000 supporting jobs.[68]

GE Energy and Orion New Zealand Limited (Orion) have announced implementation of the first phase of a GE network management system to help improve power reliability for customers. GE's ENMAC Distribution Management System is the foundation of Orion's initiative. The system of smart grid technologies will significantly improve the network company's ability to manage big network emergencies and help it to restore power faster when outages occur.

GE unveiled a 40W replacement Energy Smart LED bulb, to be available late 2010 or early 2011. The company claims that the new LED bulb will provide a 77% energy savings and produce nearly the same light output as a 40W incandescent bulb, while lasting more than 25 times as long.[69]

Educational initiatives

GE Healthcare is collaborating with The Wayne State University School of Medicine and the Medical University of South Carolina to offer an integrated radiology curriculum during their respective MD Programs led by investigators of the Advanced Diagnostic Ultrasound in Microgravity study.[70] GE has donated over one million dollars of Logiq E Ultrasound equipment to these two institutions.[71]

Legal issues

On August 4, 2009 the SEC fined General Electric $50 million for breaking accounting rules in two separate cases, misleading investors into believing GE would meet or beat earnings expectations.[72]

GE has faced criminal action regarding its defense related operations. GE was convicted in 1990 of defrauding the US Department of Defense, and again in 1992 on charges of corrupt practices in the sale of jet engines to Israel.[73]

Notable appearances in media

GE was the focus of a 1991 short subject Academy Award-winning documentary, *Deadly Deception: General Electric, Nuclear Weapons, and Our Environment* [74],[75] that juxtaposed GE's "We Bring Good Things To Life" commercials with the true stories of workers and neighbors whose lives have been affected by the company's activities involving nuclear weapons.[76]

In the early 1950s Kurt Vonnegut was a writer for General Electric. A number of his novels and stories (notably *Cat's Cradle*) refer to the fictional city of Ilium, which appears to be loosely based on Schenectady, New York. The Ilium Works is the setting for the short story "Deer in the Works".

The NBC sitcom *30 Rock* is set at the NBC Studios in the GE Building at 30 Rockefeller Center (hence the title). All characters in the main cast are NBC (and therefore GE) employees, and one character, Jack Donaghy (Alec Baldwin) is portrayed as having risen through the ranks of GE management to become Vice President of NBC's East Coast operations through the company's microwave oven division.

See also

- Edison Engineering Development Program
- GE Aviation
- GE Building
- GE Capital Finance
- GE Energy
- GE Home & Business Solutions
- GE Global Research
- GE Technology Infrastructure
- Knolls Atomic Power Laboratory, Schenectady, NY
- List of assets owned by General Electric
- Phoebus cartel
- Top 100 US Federal Contractors

References

[1] http://www.nyse.com/about/listed/quickquote.html?ticker=ge
[2] http://www.nasdaq.com/symbol/ge
[3] "GE Fact Sheet: Businesses, Locations, Awards, Leadership" (http://www.ge.com/company/factsheets/corporate.html). *GE company website*. . Retrieved May 25, 2011.
[4] "2010 Form 10-K, General Electric Company" (http://www.ge.com/investors/financial_reporting/index.html). *General Electric*. .
[5] "NBC Universal Transaction – A Comcast-GE Joint Venture" (http://www.comcast.com/nbcutransaction/). *Comcast*. .
[6] http://www.ge.com/
[7] "Company Search, EDGAR System, Securities and Exchange Commission" (http://www.sec.gov/cgi-bin/browse-edgar?action=getcompany&CIK=0000040545&owner=include&count=40). . Retrieved December 22, 2009.
[8] "GE Fact Sheet: Businesses, Locations, Awards, Leadership" (http://www.ge.com/company/factsheets/corporate.html). General Electric. . Retrieved May 10, 2011.
[9] Wald, Matthew L.. "General Electric Company" (http://topics.nytimes.com/top/news/business/companies/general_electric_company/index.html). *The New York Times*. . Retrieved May 10, 2011.
[10] "Fortune 500: GE" (http://money.cnn.com/magazines/fortune/fortune500/2011/snapshots/170.html). Fortune. 2011. . Retrieved December 17, 2010.
[11] "Fortune 20 most profitable companies: IBM" (http://money.cnn.com/galleries/2011/fortune/1104/gallery.fortune500_most_profitable.fortune/14.html). Fortune. 2011. . Retrieved December 17, 2010.
[12] "GE rankings" (http://rankingthebrands.com/Brand-detail.aspx?brandID=82). Ranking the Brands. . Retrieved December 17, 2010.
[13] "Thomas Edison & GE" (http://www.ge.com/company/history/edison.html). GE company web site. . Retrieved December 22, 2009.
[14] "History of the Dow – Timeline of Companies" (http://www.quasimodos.com/info/dowhistory.html). Quasimodos.com. January 4, 1984. . Retrieved April 23, 2012.
[15] Mahon, Morgan E. *A Flick of the Switch 1930–1950* (Antiques Electronics Supply, 1990), p.86.
[16] "GE Completes Enron Wind Acquisition; Launches GE Wind Energy" (http://findarticles.com/p/articles/mi_m0EIN/is_2002_May_10/ai_85689772/). Business Wire. May 10, 2002. . Retrieved August 9, 2010.
[17] Fairly, Peter. The Greening of GE (http://spectrum.ieee.org/energy/environment/the-greening-of-ge) *IEEE Spectrum*, July 2005. Retrieved: November 6, 2010.
[18] Arnott, Sarah (March 26, 2010). "GE to build £99m UK wind turbine plant" (http://www.independent.co.uk/news/business/news/ge-to-build-16399m-uk-wind-turbine-plant-1928159.html). *The Independent* (UK). . Retrieved December 20, 2010.
[19] Global Corruption Report 2009, Corruption and the Private Sector (http://www.transparency.org/publications/publications/global_corruption_report/gcr2009) Transparency International 2009 page 148 (English)
[20] "Kidder, Peabody & Co." (http://en.wikipedia.org/wiki/Kidder,_Peabody_&_Co.). *Wikipedia*. .
[21] Murphy, Dennis. GE completes Enron Wind acquisition; Launches GE Wind Energy (http://www.desertskywind.com/news05022002.htm) *Desert Sky Wind Farm*, May 10, 2002. Retrieved: May 1, 2010.
[22] Goldman, Davis; Pepitone, Julianne (December 3, 2009). "GE, Comcast announce joint NBC deal" (http://money.cnn.com/2009/12/03/news/companies/comcast_nbc/index.htm). CNNMoney.com. . Retrieved December 22, 2009.
[23] "Vivendi to Sell its Stake in NBCUniversal for US$5.8 billion" (http://www.vivendi.fr/vivendi/Nouvel-article,5902). Vivendi SA. December 3, 2009. . Retrieved December 22, 2009.
[24] "Vivendi To Sell 20% Stake In NBCU To GE For $5.8 Bln" (http://www.bloomberg.com/apps/news?pid=newsarchive&sid=aMoJZHPclWnY). *The Wall Street Journal*. December 3, 2009. . Retrieved December 22, 2009.

[25] "General Electric to sell stake in Garanti Bank" (http://www.newstatesman.com/banking-and-insurance/2010/03/garanti-bank-turkey-stake). March 1, 2010. . Retrieved March 2, 2010.
[26] "GE to get its heart tech from Israel" (http://www.zdnet.com/blog/healthcare/ge-to-get-its-heart-tech-from-israel/3950). August 1, 2010. . Retrieved August 21, 2010.
[27] "— GE posts sales slump, rattles recovery hopes" (http://www.comcast.net/articles/news-finance/20101006/BUSINESS-US-GE/print/). Comcast.net. . Retrieved April 23, 2012.
[28] "GE Acquires Opal Software — Strengthens Smart Grid Software Portfolio" (http://www.genewscenter.com/content/Detail.aspx?ReleaseID=11178&NewsAreaID=2). GE. . Retrieved October 19, 2010.
[29] NewsDaily: GE to buy UK oil pipe maker Wellstream for $1.3 billion reuters.com (http://www.reuters.com/article/idUSTRE6BC0U920101213)
[30] Post a Job (February 14, 2011). "Businessweek.com" (http://www.businessweek.com/news/2011-02-14/ge-agrees-to-buy-2-8-billion-oil-service-unit-shares-surge.html). *Bloomberg BusinessWeek*. . Retrieved April 23, 2012.
[31] Business Wire (March 2, 2011). "Businesswire.com" (http://www.businesswire.com/news/home/20110302005457/en/GE-Energy-Completes-Acquisition-Lineage-Power). Businesswire.com. . Retrieved April 23, 2012.
[32] "GE Capital sells Mexican assets to Santander" (http://www.financenews.co.uk/fnews/ge-capital-sells-mexican-assets-to-santander/). Finance News. . Retrieved April 23, 2012.
[33] "IRS e-file Moves Forward; Successfully Executes Electronic Filing of Nation's Largest Tax Return" (http://www.irs.gov/newsroom/article/0,,id=157845,00.html) (Press release). United States Department of the Treasury— Internal Revenue Service. May 31, 2006. . Retrieved February 1, 2007.
[34] Carney, Timothy (2011-04-07) Want to know how GE paid $0 income taxes? Think green. (http://washingtonexaminer.com/blogs/beltway-confidential/2011/04/want-know-how-ge-paid-0-income-taxes-think-green), *Washington Examiner*
[35] "GE Launches Ecomagination to Develop Environmental Technologies; Company-Wide Focus on Addressing Pressing Challenges" (http://www.businesswire.com/news/home/20050509005663/en) (Press release). General Electric. May 9, 2005. . Retrieved January 15, 2007.
[36] Deutsch, Claudia (May 22, 2007). "General Electric to Sell Plastics Division" (http://www.nytimes.com/2007/05/22/business/22plastics.html?_r=1). *The New York Times*. . Retrieved December 25, 2009.
[37] "Jeff Immelt, CEO" (http://www.ge.com/company/leadership/ceo.html). *Company Biography*. General Electric. . Retrieved December 25, 2009.
[38] "Welcome to Saskrailmuseum.org" (http://www.saskrailmuseum.org/). *Contact Us*. September 11, 2008. . Retrieved October 3, 2008.
[39] "VIDEO – General Imaging's blink detection cameras" (http://www.tgdaily.com/consumer-electronics-features/35959-video-â-general-imagingâs-blink-detection-cameras). TG Daily. February 6, 2008. . Retrieved April 23, 2012.
[40] "GE X5 Black 14MP Digital Camera w/ 15x Optical Zoom, 2.7" LCD: Digital Cameras" (http://www.walmart.com/ip/GE-X5-BK/14245324). Walmart.com. . Retrieved April 23, 2012.
[41] By metzae. "GE X500 Power Pro "bridge" DSLR camera Added a New Color" (http://dandygadget.com/2011/04/19/ge-x500-power-pro-bridge-dslr-camera-added-a-new-color/). Dandy Gadget. . Retrieved December 11, 2011.
[42] "GE confirms it's exiting appliance business" (http://www.msnbc.msn.com/id/24630072). MSNBC. May 16, 2008. . Retrieved December 25, 2009.
[43] "Best Global Brands Ranking for 2010" (http://www.interbrand.com/en/best-global-brands/best-global-brands-2008/best-global-brands-2010.aspx). Interbrand. . Retrieved February 15, 2011.
[44] "Design of G.E.'s Mark 1 Nuclear Reactors Shows Weaknesses" (http://www.nytimes.com/2011/03/16/world/asia/16contain.html?_r=1&scp=2&sq=general electric&st=cse). *The New York Times*. March 15, 2011. .
[45] Kocieniewski, David (March 24, 2011). "G.E.'s Strategies Let It Avoid Taxes Altogether" (http://www.nytimes.com/2011/03/25/business/economy/25tax.html?scp=2&sq=ge&st=cse). *The New York Times*. .
[46] Portero, Ashley. "30 Major U.S. Corporations Paid More to Lobby Congress Than Income Taxes, 2008–2010" (http://www.webcitation.org/64D9GyQG0). International Business Times. Archived from the original (http://www.ibtimes.com/articles/264481/20111209/30-major-u-s-corporations-paid-lobby.htm) on December 26, 2011. . Retrieved December 26, 2011.
[47] Political Economy Research Institute Toxic 100 Corporate Toxics Information Project Technical Notes (http://web.archive.org/web/20071114225754/http://www.peri.umass.edu/Technical-Notes.264.0.html). Retrieved November 9, 2007.
[48] "Political Economy Research Institute" (http://web.archive.org/web/20070927204419/http://www.rtknet.org/new/tox100/toxic100.php?database=t1&detail=1&datype=T&reptype=a&company1=&company2=8337&chemfac=fac&advbasic=bas). Web.archive.org. September 27, 2007. . Retrieved April 23, 2012.
[49] EPA Document Lists Firms Tied to Superfund Sites (http://projects.publicintegrity.org/superfund/report.aspx?aid=849) The Center for Public Integrity
[50] The Region; G.E. Plant Accused Of Water Pollution" (http://www.nytimes.com/1983/01/21/nyregion/the-region-ge-plant-accused-of-water-pollution.html?n=Top/Reference/Times Topics/Subjects/W/Water Pollution), *The New York Times*, January 21, 1983
[51] GE agrees to $250 million Settlement to Clean Up PCBs in Housatonic River (http://www.justice.gov/opa/pr/1999/October/471enr.htm), Department of Justice news release, October 7, 1999
[52] "Hudson River PCBs" (http://www.epa.gov/hudson/). Epa.gov. . Retrieved April 23, 2012.

[53] "Historic Hudson River Cleanup to Begin After Years of Delay, But Will General Electric Finish the Job? "Under the EPA's unusual agreement with General Electric, the company could escape full responsibility for cleaning up the toxic mess it made in the Hudson River"" (http://www.nrdc.org/water/pollution/hhudson.asp). Nrdc.org. . Retrieved April 23, 2012.
[54] Old Questions Linger as New PCB Cleanup Starts (http://query.nytimes.com/gst/fullpage.html?res=9403EFD7133EF932A35756C0A9619C8B63) *The New York Times* May 1, 2007
[55] EPA issues unilateral administrative order to General Eletric Company in Rome, Georgia (http://yosemite.epa.gov/opa/admpress.nsf/e51aa292bac25b0b85257359003d925f/5592f5848c70b2858525731b00674e9c!OpenDocument) United States Environmental Protection Agency
[56] "General Electric-designed reactors in Fukushima have 23 sisters in U.S" (http://openchannel.msnbc.msn.com/_news/2011/03/13/6256121-general-electric-designed-reactors-in-fukushima-have-23-sisters-in-us). Openchannel.msnbc.msn.com. . Retrieved April 23, 2012.
[57] "Nuclear Reactor Maps: Fukushima-Daiichi" (http://nuctrans.org/Nuc_Trans/locations/daiichi/daiichi.htm). Nuctrans.org. . Retrieved April 23, 2012.
[58] "Fukushima: Mark 1 Nuclear Reactor Design Caused GE Scientist To Quit In Protest" (http://abcnews.go.com/Blotter/fukushima-mark-nuclear-reactor-design-caused-ge-scientist/story?id=13141287). USA: ABC. March 15, 2011. . Retrieved April 23, 2012.
[59] Martin LaMonica, CNET. " GE invests in eSolar for combined solar, gas plants (http://news.cnet.com/8301-11128_3-20069472-54/ge-invests-in-esolar-for-combined-solar-gas-plants/?part=rss&subj=news&tag=2547-1_3-0-20)." Jun 6, 2011. Retrieved Jun 6, 2011.
[60] Candace Lombardi, CNET. " Solar power carport charges cars in Connecticut (http://news.cnet.com/8301-11128_3-20066460-54.html?part=rss&subj=news&tag=2547-1_3-0-20)." May 26, 2011. Retrieved May 26, 2011.
[61] "Ecomagination: Inside GE's Power Play" (http://www.worldchanging.com/archives/002669.html). Worldchanging.com. 1999-02-22. . Retrieved 2012-04-23.
[62] "Talking Green, Acting Dirty." *The New York Times* June 12, 2005 (http://www.nytimes.com/2005/06/12/opinion/nyregionopinions/12WEsullivan.html)
[63] "GE cleantech sales to top $17 billion this year" (http://www.businessgreen.com/bg/news/1804493/ge-cleantech-sales-usd17bn). Businessgreen.com. October 27, 2008. . Retrieved April 23, 2012.
[64] VentureBeat, Camille Ricketts. " GE Pumps 10B More Into Green Technology R&D (http://venturebeat.com/2010/06/24/ge-pumps-10b-more-into-green-technology-rd/)." June 24, 2010.
[65] "GE Energy Acquires Kelman Limited" (http://www.renewableenergyworld.com/rea/news/article/2008/08/ge-energy-acquires-kelman-limited-53356). Renewableenergyworld.com. August 18, 2008. . Retrieved April 23, 2012.
[66] GE to create 104 jobs in Lisburn (http://www.insideireland.ie/index.cfm/section/news/ext/generalelectric001/category/1084)
[67] GE's ecomagination™ (http://www.gepower.com/about/press/en/2008_press/120208e.htm) Certified Clarke Energy Jenbacher Power Plant Commissioned at Australian Coal Mine to Help Reduce Greenhouse Gas Emissions, 2008, www.gepower.com, Accessed April 14, 2011
[68] "GE illustrates broad spectrum of alternative energy projects" (http://www.reliableplant.com/Read/17126/ge-illustrates-broad-spectrum-of-alternative-energy-projects). Reliableplant.com. . Retrieved April 23, 2012.
[69] by Martin LaMonica (April 8, 2010). "GE makes LED replacement for 40-watt bulb | Green Tech – CNET News" (http://news.cnet.com/8301-11128_3-20002001-54.html). News.cnet.com. . Retrieved April 23, 2012.
[70] Jultrasoundmet.org (http://www.jultrasoundmed.org/cgi/content/abstract/27/5/745), A Pilot Study of Comprehensive Ultrasound Education at the Wayne State University School of Medicine
[71] (http://www.gehealthcare.com/usen/ultrasound/genimg/products/logiq_e/hryfrdhosp_waynestuni.html)
[72] Henry, David (August 4, 2009). "SEC Fines GE $50 million for Accounting Misdeeds" (http://www.businessweek.com/bwdaily/dnflash/content/aug2009/db2009084_567813.htm). *Bloomberg BusinessWeek*. . Retrieved April 23, 2012.
[73] Stevenson, Richard W. G.E. Guilty Plea in U.S. Aid to Israel (http://www.nytimes.com/1992/07/23/business/company-news-ge-guilty-plea-in-us-aid-to-israel.html), New York Times, July 23, 1992.
[74] http://www.newday.com/films/DeadlyDeception.html
[75] "Deadly Deception: General Electric, Nuclear Weapons, and Our Environment" (http://www.newday.com/films/DeadlyDeception.html). Newday.com. . Retrieved April 23, 2012.
[76] "Nuclear Weaponmakers Campaign – Corporate Accountability International – Challenging Abuse, Protecting People – Think Outside the Bottle – challenging the bottled water industry" (http://web.archive.org/web/20060621082431/http://stopcorporateabuse.org/cms/page1127.cfm). Web.archive.org. June 21, 2006. . Retrieved April 23, 2012.

Further reading

- Carlson, W. Bernard. *Innovation as a Social Process: Elihu Thomson and the Rise of General Electric, 1870–1900* (Cambridge: Cambridge University Press, 1991).
- Woodbury, David O. *Elihu Thomson, Beloved Scientist* (Boston: Museum of Science, 1944)
- Haney, John L. *The Elihu Thomson Collection* American Philosophical Society Yearbook 1944.
- Hammond, John W. *Men and Volts: The Story of General Electric*, published 1941, 436 pages.
- Mill, John M. *Men and Volts at War: The Story of General Electric in World War II*, published 1947.

External links

- General Electric corporate page (http://www.ge.com/)
- Official site for GE Appliances, the corporation's leading consumer product (http://www.geappliances.com/)
- SEC filings including 10-k (http://www.sec.gov/cgi-bin/browse-edgar?action=getcompany&CIK=0000040545&owner=exclude)
- Official site for SABIC Innovative Plastics (http://www.sabic-ip.com/).
- Historic Development of the GE.com website (http://www.vb.com/ge.htm).
- Financial Times article on GE's effort to go green (http://news.ft.com/cms/s/ac5c3e90-e60b-11da-b309-0000779e2340.html)
- GE energy (http://www.gepower.com/)
- G.E. to Double Investments in Renewables to $4B (http://www.greenbiz.com/news/2007/06/04/ge-double-investments-renewables-4b)
- USA Today article about GE's success at creating CEOs at large companies (http://www.usatoday.com/money/companies/management/2008-01-08-ceo-companies_N.htm)

Video clips

- GE Reports YouTube channel (http://www.youtube.com/user/GEreports)

Business data

- General Electric (http://finance.google.com/finance?q=GE) at Google Finance
- General Electric (http://finance.yahoo.com/q?s=GE) at Yahoo! Finance
- General Electric (http://www.hoovers.com//--ID__10634--/free-co-factsheet.xhtml) at Hoover's
- General Electric (http://www.reuters.com/finance/stocks/overview?symbol=GE) at Reuters
- General Electric (http://google.brand.edgar-online.com/?sym=GE) SEC filings at EDGAR Online
- General Electric (http://www.sec.gov/cgi-bin/browse-edgar?action=getcompany&CIK=40545) SEC filings at the Securities and Exchange Commission

Application-specific_integrated_circuit

An **application-specific integrated circuit** (ASIC) (pronounced English pronunciation: /ˈeɪsɪk/) is an integrated circuit (IC) customized for a particular use, rather than intended for general-purpose use. For example, a chip designed to run in a digital voice recorder is an ASIC. Application-specific standard products (ASSPs) are intermediate between ASICs and industry standard integrated circuits like the 7400 or the 4000 series.

A tray of **Application-specific integrated circuit (ASIC)** chips

As feature sizes have shrunk and design tools improved over the years, the maximum complexity (and hence functionality) possible in an ASIC has grown from 5,000 gates to over 100 million. Modern ASICs often include entire 32-bit processors, memory blocks including ROM, RAM, EEPROM, Flash and other large building blocks. Such an ASIC is often termed a SoC (system-on-chip). Designers of digital ASICs use a hardware description language (HDL), such as Verilog or VHDL, to describe the functionality of ASICs.

Field-programmable gate arrays (FPGA) are the modern-day technology for building a breadboard or prototype from standard parts; programmable logic blocks and programmable interconnects allow the same FPGA to be used in many different applications. For smaller designs and/or lower production volumes, FPGAs may be more cost effective than an ASIC design even in production. The non-recurring engineering (NRE) cost of an ASIC can run into the millions of dollars.

History

The initial ASICs used gate array technology. Ferranti produced perhaps the first gate-array, the ULA (Uncommitted Logic Array), around 1980. An early successful commercial application was the ULA circuitry found in the 8-bit ZX81 and ZX Spectrum low-end personal computers, introduced in 1981 and 1982. These were used by Sinclair Research (UK) essentially as a low-cost I/O solution aimed at handling the computer's graphics. Some versions of ZX81/Timex Sinclair 1000 used just four chips (ULA, 2Kx8 RAM, 8Kx8 ROM, Z80A CPU) to implement an entire mass-market personal computer with built-in BASIC interpreter.

Customization occurred by varying the metal interconnect mask. ULAs had complexities of up to a few thousand gates. Later versions became more generalized, with different base dies customised by both metal and polysilicon layers. Some base dies include RAM elements.

Standard-cell design

In the mid 1980s, a designer would choose an ASIC manufacturer and implement their design using the design tools available from the manufacturer. While third-party design tools were available, there was not an effective link from the third-party design tools to the layout and actual semiconductor process performance characteristics of the various ASIC manufacturers. Most designers ended up using factory-specific tools to complete the implementation of their designs. A solution to this problem, which also yielded a much higher density device, was the implementation of standard cells. Every ASIC manufacturer could create functional blocks with known electrical characteristics, such as propagation delay, capacitance and inductance, that could also be represented in third-party tools. Standard-cell design is the utilization of these functional blocks to achieve very high gate density and good electrical performance. Standard-cell design fits between Gate Array and Full Custom design in terms of both its non-recurring engineering and recurring component cost.

By the late 1990s, logic synthesis tools became available. Such tools could compile HDL descriptions into a gate-level netlist. Standard-cell Integrated Circuits (ICs) are designed in the following conceptual stages, although these stages overlap significantly in practice.

1. A team of design engineers starts with a non-formal understanding of the required functions for a new ASIC, usually derived from Requirements analysis.
2. The design team constructs a description of an ASIC to achieve these goals using an HDL. This process is analogous to writing a computer program in a high-level language. This is usually called the RTL (Register transfer level) design.
3. Suitability for purpose is verified by functional verification. This may include such techniques as logic simulation, formal verification, emulation, or creating an equivalent pure software model (see Simics, for example). Each technique has advantages and disadvantages, and often several methods are used.
4. Logic synthesis transforms the RTL design into a large collection of lower-level constructs called standard cells. These constructs are taken from a standard-cell library consisting of pre-characterized collections of gates (such as 2 input nor, 2 input nand, inverters, etc.). The standard cells are typically specific to the planned manufacturer of the ASIC. The resulting collection of standard cells, plus the needed electrical connections between them, is called a gate-level netlist.
5. The gate-level netlist is next processed by a placement tool which places the standard cells onto a region representing the final ASIC. It attempts to find a placement of the standard cells, subject to a variety of specified constraints.
6. The routing tool takes the physical placement of the standard cells and uses the netlist to create the electrical connections between them. Since the search space is large, this process will produce a "sufficient" rather than "globally optimal" solution. The output is a file which can be used to create a set of photomasks enabling a semiconductor fabrication facility (commonly called a 'fab') to produce physical ICs.
7. Given the final layout, circuit extraction computes the parasitic resistances and capacitances. In the case of a digital circuit, this will then be further mapped into delay information, from which the circuit performance can be estimated, usually by static timing analysis. This, and other final tests such as design rule checking and power analysis (collectively called signoff) are intended to ensure that the device will function correctly over all extremes of the process, voltage and temperature. When this testing is complete the photomask information is released for chip fabrication.

These steps, implemented with a level of skill common in the industry, almost always produce a final device that correctly implements the original design, unless flaws are later introduced by the physical fabrication process.

The design steps (or flow) are also common to standard product design. The significant difference is that standard-cell design uses the manufacturer's cell libraries that have been used in potentially hundreds of other design implementations and therefore are of much lower risk than full custom design. Standard cells produce a design density that is cost effective, and they can also integrate IP cores and SRAM (Static Random Access Memory) effectively, unlike Gate Arrays.

Gate-array design

Gate-array design is a manufacturing method in which the diffused layers, i.e. transistors and other active devices, are predefined and wafers containing such devices are held in stock prior to metallization—in other words, unconnected. The physical design process then defines the interconnections of the final device. For most ASIC manufacturers, this consists of from two to as many as nine metal layers, each metal layer running perpendicular to the one below it. Non-recurring engineering costs are much lower, as photolithographic masks are required only for the metal layers, and production cycles are much shorter, as metallization is a comparatively quick process.

Gate-array ASICs are always a compromise as mapping a given design onto what a manufacturer held as a stock wafer never gives 100% utilization. Often difficulties in routing the interconnect require migration onto a larger array device with consequent increase in the piece part price. These difficulties are often a result of the layout software used to develop the interconnect.

Pure, logic-only gate-array design is rarely implemented by circuit designers today, having been replaced almost entirely by field-programmable devices, such as field-programmable gate arrays (FPGAs), which can be programmed by the user and thus offer minimal tooling charges non-recurring engineering, only marginally increased piece part cost, and comparable performance. Today, gate arrays are evolving into structured ASICs that consist of a large IP core like a CPU, DSP unit, peripherals, standard interfaces, integrated memories SRAM, and a block of reconfigurable, uncommited logic. This shift is largely because ASIC devices are capable of integrating such large blocks of system functionality and "system-on-a-chip" requires far more than just logic blocks.

In their frequent usages in the field, the terms "gate array" and "semi-custom" are synonymous. Process engineers more commonly use the term "semi-custom", while "gate-array" is more commonly used by logic (or gate-level) designers.

Full-custom design

By contrast, full-custom ASIC design defines all the photolithographic layers of the device. Full-custom design is used for both ASIC design and for standard product design.

The benefits of full-custom design usually include reduced area (and therefore recurring component cost), performance improvements, and also the ability to integrate analog components and other pre-designed — and thus fully verified — components, such as microprocessor cores that form a system-on-chip.

The disadvantages of full-custom design can include increased manufacturing and design time, increased non-recurring engineering costs, more complexity in the computer-aided design (CAD) system, and a much higher skill requirement on the part of the design team.

For digital-only designs, however, "standard-cell" cell libraries, together with modern CAD systems, can offer considerable performance/cost benefits with low risk. Automated layout tools are quick and easy to use and also offer the possibility to "hand-tweak" or manually optimize any performance-limiting aspect of the design.

Structured design

Structured ASIC design (also referred to as "platform ASIC design"), is a relatively new term in the industry, resulting in some variation in its definition. However, the basic premise of a structured ASIC is that both manufacturing cycle time and design cycle time are reduced compared to cell-based ASIC, by virtue of there being pre-defined metal layers (thus reducing manufacturing time) and pre-characterization of what is on the silicon (thus reducing design cycle time). One definition states that

> In a "structured ASIC" design, the logic mask-layers of a device are predefined by the ASIC vendor (or in some cases by a third party). Design differentiation and customization is achieved by creating custom metal layers that create custom connections between predefined lower-layer logic elements. "Structured ASIC" technology is seen as bridging the gap between field-programmable gate arrays and "standard-cell" ASIC designs. Because only a small number of chip layers must be custom-produced, "structured ASIC" designs have much smaller non-recurring expenditures (NRE) than "standard-cell" or "full-custom" chips, which require that a full mask set be produced for every design.

This is effectively the same definition as a gate array. What makes a structured ASIC different is that in a gate array, the predefined metal layers serve to make manufacturing turnaround faster. In a structured ASIC, the use of predefined metallization is primarily to reduce cost of the mask sets as well as making the design cycle time significantly shorter. For example, in a cell-based or gate-array design the user must often design power, clock, and test structures themselves; these are predefined in most structured ASICs and therefore can save time and expense for the designer compared to gate-array. Likewise, the design tools used for structured ASIC can be substantially lower cost and easier (faster) to use than cell-based tools, because they do not have to perform all the functions that cell-based tools do. In some cases, the structured ASIC vendor requires that customized tools for their device (e.g., custom physical synthesis) be used, also allowing for the design to be brought into manufacturing more quickly.

One other important aspect about structured ASIC is that it allows intellectual property (IP) that is common to certain applications or industry segments to be "built in", rather than "designed in". By building the IP directly into the architecture the designer can again save both time and money compared to designing IP into a cell-based ASIC.

Cell libraries, IP-based design, hard and soft macros

Cell libraries of logical primitives are usually provided by the device manufacturer as part of the service. Although they will incur no additional cost, their release will be covered by the terms of a non-disclosure agreement (NDA) and they will be regarded as intellectual property by the manufacturer. Usually their physical design will be pre-defined so they could be termed "hard macros".

What most engineers understand as "intellectual property" are IP cores, designs purchased from a third-party as sub-components of a larger ASIC. They may be provided as an HDL description (often termed a "soft macro"), or as a fully routed design that could be printed directly onto an ASIC's mask (often termed a hard macro). Many organizations now sell such pre-designed cores — CPUs, Ethernet, USB or telephone interfaces — and larger organizations may have an entire department or division to produce cores for the rest of the organization. Indeed, the wide range of functions now available is a significant factor in the phenomenal increase in electronics in the late 1990s and early 2000s; as a core takes a lot of time and investment to create, its re-use and further development cuts product cycle times dramatically and creates better products. Additionally, organizations such as OpenCores are collecting free IP cores paralleling the open source movement in software.

Soft macros are often process-independent, i.e., they can be fabricated on a wide range of manufacturing processes and different manufacturers. Hard macros are process-limited and usually further design effort must be invested to migrate (port) to a different process or manufacturer.

Multi-project wafers

Some manufacturers offer Multi-Project Wafers (MPW) as a method of obtaining low cost prototypes. Often called shuttles, these MPW, containing several designs, run at regular, scheduled intervals on a "cut and go" basis, usually with very little liability on the part of the manufacturer. The contract involves the assembly and packaging of a handful of devices. The service usually involves the supply of a physical design data base i.e. masking information or Pattern Generation (PG) tape. The manufacturer is often referred to as a "silicon foundry" due to the low involvement it has in the process.

ASIC suppliers

There are two different types of ASIC suppliers, IDM and fabless. An IDM supplier's ASIC product is based in large part on proprietary technology such as design tools, IP, packaging, and usually although not necessarily the process technology. Fabless ASIC suppliers rely almost exclusively on outside suppliers for their technology. The classification can be confusing since several IDM's are also fabless semiconductor companies.

IDM ASIC suppliers

- Avago Technologies
- Elmos Semiconductor
- Cavium Networks
- Fujitsu
- Freescale Semiconductor
- HITACHI
- Global Foundries [1]
- IBM
- Infineon Technologies
- GigOptix (ChipX Structured ASIC) [2]
- LSI Corporation
- Maxim Integrated Products
- NEC
- NXP Semiconductors
- ON Semiconductor
- Renesas
- Samsung
- STMicroelectronics
- Texas Instruments
- Toshiba
- TSMC
- Bosch

Fabless ASIC suppliers

- Alchip
- Aeroflex Colorado Springs
- Brite Semiconductor
- Broadcom Corporation
- Custom Silicon Solutions, Inc. [3]
- GigOptix (ChipX Structured ASIC) [2]
- DELTA Microelectronics
- eASIC
- eSilicon
- Faraday Technology
- Hong Kong Science and Technology Parks Corporation
- ICsense (ASIC Design and Supply) [4]
- Marvell Semiconductor
- MOSIS
- Nvidia

- Open-Silicon
- PMC Sierra
- Qualcomm
- SWINDON Silicon Systems
- Socle
- Triad Semiconductor (Configurable Analog and Mixed-Signal) [5]
- Xilinx
- Cactus Semiconductor

See also

- Complex programmable logic device (CPLD)
- Electronic design automation
- Field-programmable gate array (FPGA)
- Multi Project Chip
- Very-large-scale integration (VLSI)
- System-on-a-chip (SoC)

References

- Kevin Morris (23 November 2003). "Cost-Reduction Quagmire: Structured ASIC and Other Options" [6]. *FPGA and Programmable Logic Journal.*
- Anthony Cataldo (26 March 2002). "Xilinx looks to ease path to custom FPGAs" [7]. *EE Times* (CMP Media, LLC).
- "Xilinx intros next-gen EasyPath FPGAs priced below structured ASICs" [8]. *EDP Weekly's IT Monitor* (Millin Publishing, Inc.). 18 October 2004.
- Golshan, K. (2007). *Physical design essentials: an ASIC design implementation perspective.* New York: Springer. ISBN 0 387 36642 3.

References

[1] http://www.globalfoundries.com/
[2] http://www.gigoptix.com
[3] http://www.customsiliconsolutions.com
[4] http://www.icsense.com
[5] http://www.triadsemi.com
[6] http://www.fpgajournal.com/articles/20041123_quagmire.htm
[7] http://www.eetimes.com/story/OEG20020325S0060
[8] http://findarticles.com/p/articles/mi_m0GZQ/is_41_45/ai_n8968679

Digital_Equipment_Corporation

Industry	Computer manufacturing
Fate	Assets were sold to various companies. What remained was sold to Compaq.
Successor(s)	Hewlett-Packard (2002 – present) Compaq (1998–2002)
Founded	1957
Defunct	1998
Headquarters	Maynard, Massachusetts, United States
Key people	Ken Olsen (founder, president, and chairman) Harlan Anderson (co-founder)
Products	PDP minicomputers VAX minicomputers Alpha servers and workstations DECnet VT100 terminal StrongARM microprocessors Digital Linear Tape
Employees	over 140,000 (1987)

Digital Equipment Corporation was a major American company in the computer industry and a leading vendor of computer systems, software and peripherals from the 1960s to the 1990s. Also known as **DEC**[1] and using the trademark **DIGITAL**, its PDP and VAX products were the most successful (in terms of sales) minicomputers.

From 1957 until 1992 its headquarters was located in a former wool mill at Clock Tower Place, Maynard, Massachusetts. DEC was acquired in June 1998 by Compaq, which subsequently merged with Hewlett-Packard in May 2002. Some parts of DEC, notably the compiler business and the Hudson, Massachusetts facility, were sold to Intel.

Digital Equipment Corporation should not be confused with the unrelated companies Digital Research, Inc or Western Digital (despite the latter manufacturing the LSI-11 chipsets used in DEC's low end PDP-11/03 computers).

Overview

Initially focusing on the small-end of the computer market allowed DEC to grow without its potential competitors making serious efforts to compete with them. Their PDP series of machines became popular in the 1960s, especially the PDP-8, widely considered to be the first successful minicomputer. Looking to simplify and update their line, DEC replaced most of their smaller machines with the PDP-11 in 1970, eventually selling over 600,000 units and cementing DECs position in the industry. Originally designed as a follow-on to the PDP-11, DEC's VAX-11 series was the first widely used 32-bit minicomputer, sometimes referred to as "superminis". These were able to compete in many roles with larger mainframe computers, such as the IBM System/370. The VAX was a best-seller, with over 400,000 sold, and its sales through the 1980s propelled the company into the second largest in the industry. At its peak, DEC was the second largest employer in Massachusetts, second only to the state government.

The rapid rise of the business microcomputer in the late 1980s, and especially the introduction of powerful 32-bit systems in the 1990s, quickly eroded the value of DEC's systems. DEC's last major attempt to find a space in the rapidly changing market was the DEC Alpha 64-bit RISC processor architecture. DEC initially started work on Alpha as a way to re-implement their VAX series, but also employed it in a range of high-performance workstations. Although the Alpha processor family met both of these goals, and, for most of its lifetime, was the fastest processor family on the market, it did little to affect the bottom line or repair the company's status.

The company was acquired in June 1998 by Compaq, in what was at that time the largest merger in the history of the computer industry. At the time, Compaq was focused on the enterprise market and had recently purchased several other large vendors. DEC was a major player overseas where Compaq had less presence. However, Compaq had little idea what to do with its acquisitions, and soon found itself in financial difficulty of its own. The company subsequently merged with Hewlett-Packard in May 2002. As of 2007 some of DEC's product lines were still produced under the HP name.

History

Origins

Ken Olsen and Harlan Anderson were two engineers who had been working at MIT Lincoln Laboratory on the lab's various computer projects. The Lab is best known for their work on what would today be known as "interactivity", and their machines were among the first where operators had direct control over programs running in real-time. These had started in 1944 with the famed Whirlwind which was originally developed to make a flight simulator for the US Navy, although this was never completed.[2] Instead, this effort evolved into the SAGE system for the US Air Force, which used large screens and light guns to allow operators to interact with radar data stored in the computer.[3]

DEC was headquartered at a former wool mill at Clock Tower Place, Maynard MA from 1957 until 1992

When the Air Force project wound down, the Lab turned their attention to an effort to build a version of the Whirlwind using transistors in place of vacuum tubes. In order to test their new circuitry, they first built a small 18-bit machine known as TX-0 which first ran in 1956.[4] When the TX-0 successfully proved the basic concepts, attention turned to a much larger system, the 36-bit TX-2 with a then-enormous 64 kWords of core memory. Core was so expensive that parts of TX-0's memory were stripped for the TX-2, and what remained of the TX-0 was then given to MIT on permanent loan.[5]

At MIT, Olsen and Anderson noticed something odd: students would line up for hours to get a turn to use the stripped-down TX-0, while largely ignoring a faster IBM machine that was also available. The two decided that the draw of interactive computing was so strong that they felt there was a market for a small machine dedicated to this role, essentially a commercialized TX-0. They could sell this to users where graphical output or realtime operation would be more important than outright performance. Additionally, as the machine would cost much less than the larger systems then available, it would also be able to serve users that needed a lower-cost solution dedicated to a specific task, where a larger 36-bit machine wouldn't be needed.[6]

In 1957 when the pair and Ken's brother Stan went looking for capital, they found that the American business community was hostile to investing in computer companies. Many smaller computer companies had come and gone in the 1950s, wiped out when new technical developments rendered their platforms obsolete, and even large companies like RCA and General Electric were failing to make a profit in the market. The only serious expression of interest came from Georges Doriot and his American Research and Development Corporation (AR&D). Worried that a new computer company would find it difficult to arrange further financing, Doriot suggested the fledgling company change its business plan to focus less on computers, and even change their name from "Digital Computer Corporation".[6]

The pair returned with an updated business plan that outlined two-phases for the company's development. They would start by selling computer modules as stand-alone devices that could be purchased separately and wired together to produce a number of different digital systems for lab use. Then, if these "digital modules" were able to build a self-sustaining business, the company would be free to use them to develop a complete computer in their Phase II.[7] The newly christened DEC received $70,000 from AR&D for a 70% share of the company,[6] and began operations in a Civil War era textile mill in Maynard, Massachusetts, where plenty of inexpensive manufacturing space was available.

Digital modules

In early 1958 DEC shipped its first products, the "Digital Laboratory Module" line. The Modules consisted of a number of individual electronic components and germanium transistors mounted to a circuit board, the actual circuits being based on those from the TX-2.[8]

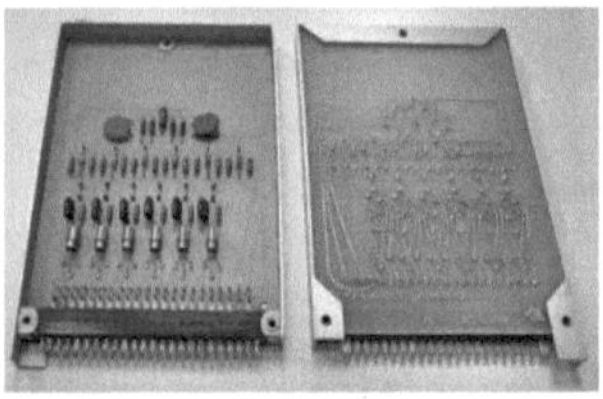

System Building Blocks 1103 hex-inverter card (both sides)

The Laboratory Modules were packaged in an extruded aluminum housing,[9] intended to sit on an engineer's workbench. They were then connected together using banana plug patch cords inserted at the front of the modules. Three versions were offered, running at 5 MHz (1957), 500 kHz (1959), or 10 MHz (1960).[8] The Modules proved to be in high demand in other computer companies, who used them to build equipment to test their own systems. Despite the recession of the late 1950s, the company sold $94,000 worth of these modules during 1958 alone, turning a profit at the end of its first year.[6]

The original Laboratory Modules were soon supplemented with the "Digital Systems Module" line, which were identical internally but packaged differently. The Systems Modules were designed with all of the connections at the back of the module using 22-pin Amphenol connectors, and were attached to each other by inserting them into a custom 19-inch rack. These versions allowed 25 modules be to inserted into a single 5-1/4 inch section of racking, and allowed the high densities needed to build a computer.[8] DEC used the Systems Modules to build their "Memory Test" machine for testing core memory systems, selling about 50 of these pre-packaged units over the next eight years.[10]

Modules were part of DEC's product line into the 1970s, although they went through several evolutions during this time as technology changed. The same circuits were then packaged as the first "R" (red) series "Flip-Chip" modules. Later, other module series provided additional speed, much higher logic density, and industrial I/O capabilities.[11] Digital published extensive data about the modules in free catalogs that became very popular.

PDP-1 family

With the company established and a successful product on the market, DEC turned its attention to the computer market once again as part of its planned "Phase II".[7] In August 1959, Ben Gurley started design of the company's first computer, the PDP-1. In keeping with Doriot's instructions, the name was an initialism for "Programmable Data Processor", leaving off the term "computer". As Gurley put it, "We aren't building computers, we're building 'Programmable Data Processors'." The prototype was first shown publicly at the Joint Computer Conference in Boston in December 1959.[12] The first PDP-1 was delivered to Bolt, Beranek and Newman in November 1960,[13] and formally accepted the next April.[14] The PDP-1 sold in basic form for $120,000, or about $900,000 in 2011 US dollars.[15] By the time production ended in 1969, 53 PDP-1s had been delivered.[10] [16]

A PDP-1 system, with Steve Russell, developer of Spacewar! at the console. This is a canonical example of the PDP-1, with the console typewriter on the left, CPU and main control panel in the center, the Type 30 display on the right.

The PDP-1 was supplied standard with 4096 words of core memory, 18-bits per word, and ran at a basic speed of 100,000 operations per second. It was constructed using many System Building Blocks that were packaged into several 19-inch racks. The racks were themselves packaged into a single large mainframe case, with a hexagonal control panel containing switches and lights mounted to lay at table-top height at one end of the mainframe. Above the control panel was the system's standard input/output solution, a punch tape reader and writer. Most systems were purchased with two peripherals, the Type 30 vector graphics display, and a Soroban Engineering modified IBM Model B Electric typewriter that was used as a printer. The Soroban system was notoriously unreliable, and often replaced with a modified Friden Flexowriter, which also contained its own punch tape system. A variety of more-expensive add-ons followed, including magnetic tape systems, punched card readers and punches, and faster punch tape and printer systems.

When DEC introduced the PDP-1, they also mentioned larger machines at 24, 30 and 36-bits, based on the same design.[17] During construction of the prototype PDP-1, some design work was carried out on a 24-bit PDP-2, and the 36-bit PDP-3. Although the PDP-2 never proceeded beyond the initial design, the PDP-3 found some interest and was designed in full.[18] Only one PDP-3 appears to have been built, in 1960, by the CIA's Scientific Engineering Institute (SEI) in Waltham, Massachusetts. According to the limited information available, they used it to process radar cross section data for the Lockheed A-12 reconnaissance aircraft. Gordon Bell remembered that it was being used in Oregon some time later, but could not recall who was using it.[19]

In November 1962 DEC introduced the $65,000 PDP-4. The PDP-4 was similar to the PDP-1 and used a similar instruction set, but used slower memory and different packaging to lower the price. Like the PDP-1, about 54 PDP-4's were eventually sold, most to a customer base similar to the original PDP-1.[20]

In 1964 DEC introduced its new Flip Chip module design, and used it to re-implement the PDP-4 as the PDP-7. The PDP-7 was introduced in December 1964, and about 120 were eventually produced.[21] An upgrade to the Flip Chip led to the R series, which in turn led to the PDP-7A in 1965.[22] The PDP-7 is most famous as the original machine for the Unix operating system.[23]

A more dramatic upgrade to the PDP-1 series was introduced in August 1966, the PDP-9.[24] The PDP-9 was instruction compatible with the PDP-4 and –7, but ran about twice as fast as the –7 and was intended to be used in larger deployments. At only $19,900 in 1968,[25] the PDP-9 was a big seller, eventually selling 445 machines, more than all of the earlier models combined.[26]

Even while the PDP-9 was being introduced, its replacement was being designed, and was introduced as 1969's PDP-15, which re-implemented the PDP-9 using integrated circuits in place of modules. Much faster than the PDP-9 even in basic form, the PDP-15 also included a floating point unit and a separate input/output processor for further performance gains. Over 400 PDP-15's were ordered in the first eight months of production, and production eventually amounted to 790 examples in 12 basic models.[26] However, by this time other machines in DEC's lineup could fill the same niche at even lower price points, and the PDP-15 would be the last of the 18-bit series.

PDP-8 family

In 1962, Lincoln Laboratory used a selection of System Building Blocks to implement a small 12-bit machine, and attached it to a variety of analog-to-digital (A to D) input/output (I/O) devices that made it easy to interface with various analog lab equipment. The LINC proved to attract intense interest in the scientific community, and has since been referred to as the first real minicomputer,[27] a machine that was small and inexpensive enough to be dedicated to a single task even in a small lab.

A PDP-8 on display at the Smithsonian's National Museum of American History in Washington, D.C.. This example is from the first generation of PDP-8s, built with discrete transistors and later known as the *Straight 8*.

Seeing the success of the LINC, in 1963 DEC took the basic logic design but stripped away the extensive A to D systems to produce the PDP-5. The new machine, the first outside the PDP-1 mould, was introduced at WESTCON on 11 August 1963. A 1964 ad expressed the main advantage of the PDP-5, "Now you can own the PDP-5 computer for what a core memory alone used to cost: $27,000"[28] 116 PDP-5s were produced until the lines were shut down in early 1967. Like the PDP-1 before it, the PDP-5 inspired a series of newer models based on the same basic design that would go on to be more famous than its parent.

On 22 March 1965, DEC introduced the PDP-8, which replaced the PDP-5's modules with the new R-series modules using Flip Chips. The machine was re-packaged into a small tabletop case, which remains distinctive for its use of smoked plastic over the CPU which allowed one to easily see the wire-wrapped internals of the CPU. Sold standard with 4 kWords of 12-bit core memory and a Teletype Model 33 ASR for basic input/output, the machine listed for only $18,000. The PDP-8 is referred to as the first *real* minicomputer because of its sub-$25,000 price.[29] [30] Sales were, unsurprisingly, very strong, and helped by the fact that several competitors had just entered the market with machines aimed directly at the PDP-5's market space, which the PDP-8 trounced. This gave the company two years of unrestricted leadership,[31] and eventually 1450 "straight eight" machines were produced before it was replaced by newer implementations of the same basic design.[28]

DEC hit an even lower price-point with the PDP-8/S, the S for "serial". As the name implies the /S used a serial arithmetic unit, which was much slower but reduced costs so much that the system sold for under $10,000.[32] DEC then used the new PDP-8 design as the basis for a new LINC, the two-processor LINC-8. The LINC-8 used one PDP-8 CPU and a separate LINC CPU, and included instructions to switch from one to the other. This allowed customers to run their existing LINC programs, or "upgrade" to the PDP-8, all in software. Although not a huge seller, 142 LINK-8s were sold starting at $38,500.[28] Like the original LINC to PDP-5 evolution, the LINC-8 was then modified into the single-processor PDP-12, adding another 1000 machines to the 12-bit family.[28] [33] Newer circuitry designs led to the PDP-8/I and PDP-8/L in 1968.[11] In 1975, one year after an agreement between Digital and Intersil, the Intersil 6100 chip was launched, effectively a PDP-8 on a chip. This was a way to allow PDP-8 software to be run even after the official end-of-life announcement for the Digital PDP-8 product line.

PDP-10 family

A "B" (blue) series Flip Chip module containing nine transistors, 1971

While the PDP-5 introduced a lower-cost line, 1963's PDP-6 was intended to take DEC into the mainframe market with a 36-bit machine. However, the PDP-6 proved to be a "hard sell" with customers, as it offered few advantages over similar machines from the better established vendors like IBM or Honeywell, in spite of its low cost around $300,000. Only 23 were sold,[34] or 26 depending on the source,[35] and unlike earlier models the low sales meant the PDP-6 was not improved with intermediate versions. However, the PDP-6 is historically important as the platform that introduced "Monitor", an early time-sharing operating system that would evolve into the widely used TOPS-10.[36]

In spite of the PDP-6's limited commercial success, it introduced many features that clearly had commercial benefit. When the Flip Chip packaging allowed the PDP-6 to be re-implemented at a much lower cost, DEC took the opportunity to carry out a similar evolution of their 36-bit design and introduced the PDP-10 in 1968. The PDP-10 was as much a success as the PDP-6 was a failure; during its lifetime about 700 mainframe PDP-10s were sold before production ended in 1984.[34] The PDP-10 was widely used in university settings, and thus was the basis of many advances in computing and operating system design during the 1970s. DEC later re-branded all of the models in the 36-bit series as the "DECsystem-10", and PDP-10s are generally referred to by the model of their CPU, like "KA10". Later upgrades produced the compatible DECSYSTEM-20, along with TOPS-20 that included virtual memory.

DECtape

One of the most unusual peripherals produced for the PDP-10 was the DECtape. The DECtape was a length of special 3/4-inch wide magnetic tape wound on 5-inch reels. The recording format was a 10-track approach using fixed-length numbered 'blocks' organized into a standard file structure, including a directory. Files could be written, read, changed, and deleted on a DECtape as though it were a disk drive. For greater efficiency, the DECtape drive could read and write to a DECtape in both directions.

In fact, some PDP-10 systems had no disks at all, using DECtapes alone for their primary data storage. The DECtape was also widely used on other PDP models, since it was much easier to use than hand-loading multiple paper tapes. Primitive early time-sharing systems could use DECtapes as system devices and swapping devices. Although superior to paper tape, DECtapes were relatively slow, and were supplanted as reliable disk drives became affordable.

PDP-11

In 1968, DEC was developing a PDP machine that would be based on 8-bit bytes instead of 6-bit characters. Known as the "PDP-X", the project was eventually cancelled. Several team members decamped and set up Data General in May 1968, and rapidly brought the 16-bit NOVA minicomputer to market. DEC immediately found itself behind in the industry transition to 8-bit bytes, and soon lost its crown as the largest minicomputer vendor.

The PDP-11 16-bit computer was designed in a crash program by Harold McFarland, Gordon Bell, Roger Cady, and others.[37] The project was able to leap forward in design with the arrival of Harold McFarland, who had been researching 16-bit designs at Carnegie Mellon University. One of his simpler designs became the PDP-11, although when they first viewed the proposal, management was not impressed and almost cancelled it.[37]

In particular, the new design did not include many of the addressing modes that were intended to make programs smaller in memory, a technique that was widely used on other DEC machines and CISC designs in general. This would mean the machine would spend more time accessing memory, which would slow it down. However, the machine also extended the idea of multiple "General Purpose Registers" (GPRs), which gave the programmer flexibility to use these high-speed memory caches as they needed, potentially addressing the performance issues.

A major advance in the PDP-11 design was Digital's Unibus, which supported all peripherals through memory mapping. This allowed a new device to be added easily, generally only requiring plugging a hardware interface board into the backplane, and then installing software that read and wrote to the mapped memory to control it. The relative ease of interfacing spawned a huge market of third party add-ons for the PDP-11, which made the machine even more useful.

The combination of architectural innovations proved superior to competitors and the "11" architecture was soon the industry leader, propelling DEC back to a strong market position. The design was later expanded to allow paged physical memory and memory protection features, useful for multitasking and time-sharing. Some models supported separate instruction and data spaces for an effective virtual address size of 128 kB within a physical address size of up to 4 MB. Smaller PDP-11s, implemented as single-chip CPUs, continued to be produced until 1996, by which time over 600,000 had been sold.[26]

The PDP-11 supported several operating systems, including Bell Labs' new Unix operating system as well as DEC's DOS-11, RSX-11, IAS, RT-11, DSM-11, and RSTS/E. Many early PDP-11 applications were developed using standalone paper-tape utilities. DOS-11 was the PDP-11's first disk operating system, but was soon supplanted by more capable systems. RSX provided a general-purpose multitasking environment and supported a wide variety of programming languages. IAS was a time-sharing version of RSX-11D. Both RSTS and Unix were time-sharing systems available to educational institutions at little or no cost, and these PDP-11 systems were destined to be the "sandbox" for a rising generation of engineers and computer scientists. Large numbers of PDP-11/70s were deployed in telecommunications and industrial control applications. AT&T became DEC's largest customer.

RT-11 provided a practical real-time operating system in minimal memory, allowing the PDP-11 to continue Digital's critical role as a computer supplier for embedded systems. Historically, RT-11 also served as the inspiration for many microcomputer OS's, as these were generally being written by programmers who cut their teeth on one of the many PDP-11 models. For example, CP/M used a command syntax similar to RT-11's, and even retained the awkward PIP program used to copy data from one computer device to another. As another historical footnote, DEC's use of "/" for "switches" (command-line options) would lead to the adoption of "\" for pathnames in MS-DOS and Microsoft Windows as opposed to "/" in Unix.[38]

The evolution of the PDP-11 followed earlier systems, eventually including a single-user deskside personal computer form, the MicroPDP-11. In total, around 600,000 PDP-11's of all models were sold. and a wide variety of third-party peripheral vendors had also entered the computer product ecosystem.

Many PDP-11-like machines were also introduced by competitors. The PDP-11 series was cloned in COMECON countries as the SM EVM series, and was produced in quantities comparable to original PDP-11 production.

VAX

In 1976, DEC decided to extend the PDP-11 architecture to 32-bits while adding a complete virtual memory system to the simple paging and memory protection of the PDP-11. The result was the VAX architecture. The first computer to use a VAX CPU was the VAX-11/780, which DEC referred to as a *superminicomputer*. Although it was not the first 32-bit minicomputer, the VAX-11/780's combination of features, price, and marketing almost immediately propelled it to a leadership position in the market after it was released in 1978. VAX systems were so successful that in 1983, DEC canceled its Jupiter project, which had been intended to build a successor to the PDP-10 mainframe, and instead focused on promoting the VAX as the single computer architecture for the company.[39]

Supporting the VAX's success was the VT52, one of the most successful smart terminals. Building on earlier less successful models (the VT05 and VT50), the VT52 was the first terminal that did everything one might want in a single chassis. The VT52 was followed by the even more successful VT100 and its follow-ons, making DEC one of the largest terminal vendors in the industry. With the VT series, DEC could now offer a complete top-to-bottom system from computer to all peripherals, which formerly required collecting the required devices from different suppliers.

The VAX processor architecture and family of systems evolved and expanded through several generations during the 1980s, culminating in the NVAX microprocessor implementation and VAX 7000/10000 series in the early 1990s.[40]

Early microcomputers

The introduction of the first general purpose microprocessors inevitably led to the first microcomputers around 1975. At the time these systems were of limited utility, and Ken Olsen famously derided them in 1977, stating "There is no reason for any individual to have a computer in his home."[41] Unsurprisingly, DEC did not put much effort into the microcomputer area in the early days of the market. At the beginning of the 1980s, DEC built the VT180 (codenamed "Robin"), which was a VT100 terminal with an added Z80-based microcomputer running CP/M, but this product was initially available only to DEC employees.[42]

It was only after IBM had successfully launched the IBM PC in 1981 that DEC responded with their own systems. In 1982, Digital introduced not one, but three incompatible machines which were each tied to different proprietary architectures. The first, the DEC Professional, was based on the PDP-11/23 (and later, the 11/73) running the RSX-11M+ derived, but menu-driven, P/OS ("Professional Operating System"). This DEC machine easily outperformed the PC, but was more expensive than, and completely incompatible with IBM PC hardware and software, offering far fewer options for customizing a system. In addition, a new user would have to learn an awkward, slow, and inflexible menu-based user interface which appeared to be radically different from PC-DOS or CP/M, which were more commonly used on the 8080 and 8088 based microcomputers of the time. A second offering, the DECmate II was the latest version of the PDP-8 based word processors, but not really suited to general computing, nor competitive with Wang Laboratories' popular word processing equipment.

The best known of DEC's early microcomputers was the Rainbow 100, which ran an implementation of CP/M on the Intel 8086 processor. Applications from standard CP/M could be re-compiled for the Rainbow, but by this time users were expecting custom built (pre compiled binary) applications such as Lotus 1 2 3, which was eventually ported along with MS-DOS 2.0 and introduced in late 1983. Although the Rainbow generated some press, it was unsuccessful due to its high price and lack of marketing and sales support.[43]

The way the DEC standard RX50[44] floppy disk drive supported DEC's initial offerings seemed to encapsulate their approach to the personal computer market. Although the mechanical drive hardware was nearly identical to other 5.25" floppy disk drives available on competing systems,[45] DEC sought to differentiate their product by using a proprietary disk format for the data written on the disk. The DEC format had a higher capacity for data, but the RX50 drives were incompatible with other PC floppy drives. This required DEC owners to buy higher-priced, specially-formatted floppy media, which was harder to obtain through standard distribution channels. DEC attempted to enforce exclusive control over its floppy media sales by copyrighting its proprietary disk format, and requiring a negotiated license agreement and royalty payments from anybody selling compatible media. The proprietary data format meant that RX50 floppies were not interchangeable with other PC floppies, further isolating DEC products from the developing de facto standard PC market. Hardware hackers and DEC enthusiasts eventually reverse-engineered the RX50 format,[44] [46] but the damage had already been done, in terms of market confusion and isolation.

A further system was introduced in 1986 as the VAXmate, which included Microsoft Windows 1.0 and used VAX/VMS-based file and print servers along with integration into DEC's own DECnet-family, providing LAN/WAN connection from PC to mainframe or supermini. The VAXmate replaced the Rainbow, and in its standard form was the first widely marketed diskless workstation.

Networking and clusters

In 1984, DEC launched its first 10 Mbit/s Ethernet. Ethernet allowed scalable networking, and VAXcluster allowed scalable computing. Combined with DECnet and Ethernet-based terminal servers (LAT), DEC had produced a networked storage architecture which allowed them to compete directly with IBM. Ethernet replaced token ring, and went on to become the dominant networking model in use today.

In September 1985, DEC became the fifth company to register a .com domain name (dec.com).

Along with the hardware and protocols, DEC also introduced the VAXcluster concept, which allowed several VAX machines to be tied together into a single larger storage system. VAXclusters allowed a DEC-based company to scale their services by adding new machines to the cluster at any time, as opposed to buying a faster machine and using that to replace a slower one. The flexibility this offered was compelling, and allowed DEC to attack high-end markets formerly out of their reach.

Diversification

Although their microcomputer efforts were eventually considered failures, the PDP-11 and VAX lines continued to sell in record numbers. Better yet, DEC was competing very well against the market leader, IBM, taking an estimated $2 billion away from them in the mid-80s. In 1986, Digital's profits rose 38 percent when the rest of the computer industry experienced a downturn, and by 1987 the company was threatening IBM's number one position in the computer industry.[6]

At its peak, Digital was the second-largest computer company in the world, with over 100,000 employees. It was during this time that the company branched out development into a wide variety of projects that were far from its core business in computer equipment. The company invested heavily in custom software. In the 1970s and earlier most software was custom-written to serve a specific task, but by the 1980s the introduction of relational databases and similar systems allowed powerful software to be built in a modular fashion, potentially saving enormous amounts of development time. Software companies like Oracle became the new darlings of the industry, and DEC started their own efforts in every "hot" niche, in some cases several projects for the same niche. Some of these products competed with DEC's own partners, notably Rdb which competed with Oracle's products on the VAX, part of a major partnership only a few years earlier.

Although many of these products were well designed, most of them were DEC-only or DEC-centric, and customers frequently ignored them and used third-party products instead. This problem was further exacerbated by Olsen's aversion to traditional advertising and his belief that well-engineered products would sell themselves. Hundreds of millions of dollars were spent on these projects, at the same time that workstations using RISC microprocessors were starting to approach VAX CPUs in performance.

Faltering in the market

As microprocessors continued to improve in the 1980s, it soon became clear that the next generation would offer performance and features equal to the best of DECs low-end minicomputer lineup. Worse, the Berkeley RISC and Stanford MIPS designs were aiming to introduce 32-bit designs that would outperform the fastest members of the VAX family, DEC's cash cow.[47]

Constrained by the huge success of their VAX/VMS products, which followed the proprietary model, the company was very late to respond to these threats. In the early 1990s, DEC found its sales faltering and its first layoffs followed. The company that created the minicomputer, a dominant networking technology, and arguably the first computers for personal use, had abandoned the "low end" market, whose dominance with the PDP-8 had built the company in a previous generation. Decisions about what to do about this threat led to infighting within the company that seriously delayed their responses.

One group suggested that every possible development in the industry be poured into the construction of a new VAX family that would leapfrog the performance of the existing machines. This would limit the market erosion in the top-end segment, where profit margins were maximized and DEC could continue to survive as a minicomputer vendor. This line of thought led, eventually, to the VAX 9000 series, which were plagued with problems when they were first introduced in October 1989, already two years late.[48] The problems took so long to work out, and the prices of the systems were so high, that DEC was never able to make the line the success they hoped.

Others within the company felt that the proper response was to introduce their own RISC designs and use those to build new machines. However, there was little official support for these efforts, and no less than four separate small projects ran in parallel at various labs around the US. Eventually these were gathered into the DEC PRISM project, which delivered a credible 32-bit design with some unique features allowing it to serve as the basis of a new VAX implementation.[49] Infighting with teams dedicated to DEC's big iron made funding difficult, and the design was not finalized until April 1988, and then cancelled shortly thereafter.[50]

Another group concluded that new workstations like those from Sun Microsystems and Silicon Graphics would take away a large part of DEC's existing customer base before the new VAX systems could address the issues, and that the company needed its own Unix workstation as soon as possible. Fed up with slow progress on both the RISC and VAX fronts, a group in Palo Alto started a skunkworks project to introduce their own systems. Selecting the MIPS processor, which was widely available, introducing the new DECstation series with the model 3100 on 11 January 1989.[51] These systems would see some success in the market, but were later displaced by similar models running the Alpha.

32-bit MIPS and 64-bit Alpha systems

Inside view of AlphaServer 2100.

Eventually, in 1992, DEC launched the DECchip 21064 processor, the first implementation of their Alpha instruction set architecture, initially named Alpha AXP (the "AXP" was a "non-acronym" and was later dropped). This was a 64-bit RISC architecture (as opposed to the 32-bit CISC architecture used in the VAX) and one of the first "pure" (not an extension of an earlier 32-bit architecture) 64-bit microprocessor architectures and implementations. The Alpha offered class-leading performance at its launch, and subsequent variants continued to do so into the 2000s. An AlphaServer SC45 supercomputer was still ranked No. 6 in the world in November 2004.[52] Alpha-based computers (the DEC AXP series, later the AlphaStation and AlphaServer series) superseded both the VAX and MIPS architecture in DEC's product lines, and could run OpenVMS, DEC OSF/1 AXP (later, Digital Unix or Tru64 UNIX) and Microsoft's then-new operating system, Windows NT.

In 1998, following the takeover by Compaq Computers, a decision was made that Microsoft would no longer support and develop Windows NT for the Alpha series computers, a decision that was seen as the beginning of the end for the Alpha series computers.

DEC tried to compete in the Unix market by adding POSIX-compatibility features to the VAX/VMS operating system (becoming "OpenVMS") and by selling its own version of Unix (Ultrix on PDP-11, VAX and MIPS architectures; OSF/1 on Alpha), and began to advertise more aggressively. DEC was simply not prepared to sell into a crowded Unix market however, and the low end PC-servers running NT (based on Intel processors) took market share from Alpha-based computers. DEC's workstation and server line never gained much popularity beyond former DEC customers.

StrongARM

In the mid-1990s, Digital Semiconductor collaborated with ARM Limited to produce the StrongARM microprocessor. This was based in part on ARM7 and in part on DEC technologies like Alpha, and was targeted at embedded systems and portable devices. It was highly compatible with the ARMv4 architecture and was very successful, competing effectively against rivals such as the SuperH and MIPS architectures in the portable digital assistant market. Microsoft subsequently dropped support for these other architectures in their Pocket PC platform. In 1997, as part of a lawsuit settlement, the StrongARM intellectual property was sold to Intel. They continued to produce StrongARM, as well as developing it into the XScale architecture. Intel subsequently sold this business to Marvell Technology Group in 2006.

Designing solutions

Beyond DECsystem-10/20, PDP, VAX and Alpha, Digital was well respected for its communication subsystem designs, such as Ethernet, DNA (Digital Network Architecture – predominantly DECnet products), DSA (Digital Storage Architecture – disks/tapes/controllers), and its "dumb terminal" subsystems including VT100 and DECserver products.[53]

Final years

New 1993 corporate logo

In June 1992, Ken Olsen was replaced by Robert Palmer as the company's president. Digital's board of directors also granted Palmer the title of chief executive officer ("CEO"), a title that had never been used during Digital's 35-year existence. Palmer had joined DEC in 1985 to run Semiconductor Engineering and Manufacturing. His relentless campaign to be CEO, and success with the Alpha microprocessor family, made him a candidate to succeed Olsen. At the same time a more modern logo was designed[54]

During the profitable years up until the early 1990s, DEC was a company that boasted that it never had a general layoff;[55] following the 1992 economic downturn, layoffs became regular events as the company continually downsized to try to stay afloat.[56] Palmer was tasked with the goal of bringing DEC back to profitability, which he attempted to do by changing the established DEC business culture, hiring new executives from outside the company, and selling off various non-core business units:[57]

- Worldwide training was spun off to form an independent/new company called Global Knowledge Network.
- Rdb, DEC's database product, was sold to Oracle.
- Rights to the PDP-11 line and several PDP-11 operating systems were sold to Mentec in 1994, though DEC continued to produce some PDP-11 hardware for a few years.[58]
- Disk and DLT technologies was sold to Quantum Corporation in 1994, immediately propelling Quantum to the No1 spot in disk manufacturing.
- Text terminal business (VT100 and its successors) was sold in August 1995 to Boundless Technologies.
- CORBA-based product, ObjectBroker, and its messaging software, MessageQ, were sold to BEA Systems, Inc in March 1997.
- In May 1997, DEC sued Intel for allegedly infringing on its Alpha patents in designing the original Pentium, Pentium Pro, and Pentium II chips.[59] As part of a settlement, much of DEC's chip design and fabrication business was sold to Intel. This included DEC's StrongARM implementation of the ARM computer architecture, which Intel marketed as the XScale processors commonly used in Pocket PCs. The core of Digital Semiconductor, the Alpha microprocessor group, remained with DEC, while the associated office buildings went to Intel as part of the Hudson, Mass. fabrication site.

- Printer business was sold in 1997 to GENICOM (now TallyGenicom), which then produced models bearing the Digital logo.
- Networking business was sold c.1997 to Cabletron Systems, and subsequently spun off as Digital Network Products Group.
- DECtalk and DECvoice voice products were spun off, and eventually arrived at Fonix.

By 1997 Digital had subsidiary companies in more than two dozen countries including Austria, Australia, Belgium, Brazil, Canada, China (People's Republic), Colombia, Cyprus, Czech Republic, Denmark, Finland, France, Germany, Ireland, Israel, Japan, Jersey States, New Zealand, Netherlands, Norway, Russia, Singapore, Spain, Sweden, Switzerland, Taiwan, and the United Kingdom.[60]

Eventually, on 26 January 1998, what remained of the company (including Digital's multivendor global services organization and customer support centers) was sold to Compaq, which was acquired by Hewlett-Packard in 2002. The remainder of Digital Semiconductor (the Alpha microprocessor group) was sold to Intel, which placed those employees back in their Hudson, Mass. office which they had vacated when the site was sold to Intel in 1997. Compaq, and later HP, continued to sell many of the former Digital products but rebranded with their own logos. For example, HP now sells what were formerly Digital's StorageWorks disk/tape products,[61] as a result of the Compaq acquisition.

The Digital logo survived for a while after the company ceased to exist, as the logo of Digital GlobalSoft, an IT services company in India (which was a 51 percent subsidiary of Compaq). Digital GlobalSoft was later renamed "HP GlobalSoft" (also known as the "HP Global Delivery India Center" or HP GDIC) and no longer uses the Digital logo.

The digital.com and DEC.com domain names are now owned by Hewlett-Packard and redirect to their US website.[62] Digital once held the Class A IP address block 16.0.0.0/8.

The Digital Federal Credit Union (DCU, now DFCU), which was chartered in 1979 for employees of DEC, is now open to essentially everyone. DFCU has over 700 different sponsors, including the companies that acquired pieces of DEC.

Research

DEC's Research Laboratories (or Research Labs, as they were commonly known) conducted Digital's corporate research. Some of them were operated by Compaq and are still operated by Hewlett-Packard. The laboratories were:

- Western Research Laboratory (WRL) in Palo Alto, California, USA
- Systems Research Center (SRC) in Palo Alto, California, USA
- Network Systems Laboratory (NSL) in Palo Alto, California, USA
- Cambridge Research Laboratory (CRL) in Cambridge, Massachusetts, USA
- Paris Research Laboratory (PRL) in Paris, France
- MetroWest Technology Campus (MTC) in Maynard, Massachusetts, USA

Some of the former employees of Digital's Research Labs or Digital's R&D in general include:

- Gordon Bell
- Leonard Bosack
- Henry Burkhardt III, co-founder of Data General Corporation and Kendall Square Research
- Mike Burrows
- Luca Cardelli
- Dave Cutler
- Ed deCastro, co-founder of Data General Corporation
- Jim Gettys
- Henri Gouraud

- Jim Gray
- Alan Kotok
- Leslie Lamport
- Butler Lampson
- Louis Monier
- Isaac Nassi
- Radia Perlman
- Marcus Ranum
- Brian Reid
- Paul Vixie

Some of the former employees of Digital Equipment Corp who were responsible for developing Alpha and StrongARM

- Daniel W. Dobberpuhl
- Jill Keller
- Rich Witek

Some of the work of the Research Labs was published in the *Digital Technical Journal*,[63] which was in published from 1985 until 1998.[64]

Accomplishments

Digital supported the ANSI standards, especially the ASCII character set, which survives in Unicode and the ISO 8859 character set family. Digital's own Multinational Character Set also had a large influence on ISO 8859-1 (Latin-1) and, by extension, Unicode.

The first versions of the C language and the Unix operating system ran on Digital's PDP series of computers (first on a PDP-7, then the PDP-11's), which were among the first commercially viable minicomputers, although for several years Digital itself did not encourage the use of Unix.

Digital also produced the popular VAX computer family, the first pure 64-bit microprocessor architecture (Alpha AXP), the first commercially successful workstation (the VT-78), and some commercially unsuccessful personal computers. The central computing system of the Soviet reusable Buran spaceship was based on two MicroVAX computers.

Digital produced widely used interactive operating systems, including OS-8, TOPS-10, TOPS-20, RSTS/E, RSX-11, RT-11, and OpenVMS. PDP computers, in particular the PDP-11 model, inspired a generation of programmers and software developers. Some PDP-11 systems more than 25 years old (software and hardware) are still being used to control and monitor factories, transportation systems and nuclear plants. Digital was an early champion of time-sharing systems.

Digital was to the command-line interface (CLI) what Apple was to the GUI : there was history before and innovation after, but it was Digital's operating systems that put it together in a complete and definitive form . The command-line interfaces found in Digital's systems, eventually codified as DCL, would look familiar to any user of modern microcomputer CLIs; those used in earlier systems, such as CTSS, IBM's JCL, or Univac's time-sharing systems, would look utterly alien. Many features of the CP/M and MS-DOS CLI show a recognizable family resemblance to Digital's OSes, including command names such as DIR and HELP and the "name-dot-extension" file naming conventions.

VAX and MicroVAX computers (very widespread in the 1980s) running VMS formed one of the most important proprietary networks, DECnet, which linked business and research facilities. The DECnet protocols formed one of the first peer-to-peer networking standards, with DECnet phase I being released in the mid-1970s. Email, file sharing, and distributed collaborative projects existed within the company long before their value was recognized in the market.

Digital, Intel and Xerox through their collaboration to create the DIX standard, were champions of Ethernet, but Digital is the company that made Ethernet commercially successful. Initially, Ethernet-based DECnet and LAT protocols interconnected VAXes with DECserver terminal servers. Starting with the UNIBUS to Ethernet adapter, multiple generations of Ethernet controllers from Digital were the de facto standard. The CI "computer interconnect" adapter was the industry's first network interface controller to use separate transmit and receive "rings".

Digital also invented clustering, an operating system technology that treated multiple machines as one logical entity. Clustering permitted sharing of pooled disk and tape storage via the HSC50/70/90 and later series of Hierarchical Storage Controllers. HSCs delivered the first hardware RAID 0 and 1 capabilities and the first serial interconnects of multiple storage technologies. This technology was the forerunner to systems like Network of Workstations which are used for massively cooperative tasks such as web-searches and drug research.

The LA36 and LA120 dot matrix printers became industry standards and may have hastened the demise of the Teletype Corporation.

The VT100 computer terminal became the industry standard, implementing a useful subset of the ANSI X3.64 standard, and even today terminal emulators such as HyperTerminal, PuTTY and Xterm still emulate a VT100 (or its more capable successor, the VT220).

The X Window System, the network transparent window system used on UNIX and Linux, and also available on other operating systems, was developed at MIT jointly between Project Athena and the Laboratory for Computer Science. Digital was the primary sponsor for this project, which was one of the first large scale free software projects, a contemporary of the GNU Project but not associated with it.

Dave Cutler, who led the development of RSX-11M, RSX-11M+, VMS and then VAXeln, left Digital in 1988 to lead the development of Windows NT which was initially intended to be an "open" operating system that would run on more powerful processors than the existing predominant x86 architecture. While DEC did see value in continuing its X Window System environment to support engineering customers, it understood that future mid-range based customers would demand a faster and more user friendly Graphical User Interface (GUI), potentially offered through the Windows environment. As Microsoft had experience, technology and patents that would give them upper hand in advancing GUI technology, DEC opted to partner with Microsoft's development of Windows NT for mid-range processors and share in the R&D process of creating faster computers using their RISC Alpha processors. Microsoft was not exclusively bound to the Alpha chip so it pursued other processor makers such as IBM with the PowerPC architecture and eventually capitalized on the emerging strength of the Intel x86 based processors. While most would say that the market determined its own course through the more cost efficient processors produced by Intel, DEC quickly came to realize that its loose partnership with Microsoft would lead to its own undoing in terms of being a major player in the mid-range processor market.

Notes-11 and its follow-on product, VAX Notes, were two of the first examples of online collaboration software, a category that has become to be known as groupware. Len Kawell, one of the original Notes-11 developers later joined Lotus Development Corporation and contributed to their Lotus Notes product.

Digital was one of the first businesses connected to the Internet with *dec.com*, registered in 1985,[65] being one of the first of the now ubiquitous *.com* domains. *gatekeeper.dec.com* was a well-known software repository during the pre-World Wide Web days, and Digital was also the first computer vendor to open a public website, on 1 October 1993.[66] The popular AltaVista, created by Digital, was one of the first comprehensive Internet search engines. (Although Lycos was earlier, it was much more limited.)

DEC invented Digital Linear Tape (DLT), formerly known as CompacTape, which began as a compact backup medium for MicroVAX systems, and later grew to capacities of 800 gigabytes.

Work on the first hard-disk-based MP3-player, the Personal Jukebox, started at the DEC Systems Research Center. (The project was started about a month before the merger into Compaq was completed.)

DEC's Western Research Lab created the Itsy Pocket Computer. This was developed into the Compaq iPaq line of PDAs, which replaced the Compaq Aero PDA.

User organizations

Originally the users' group was called DECUS (Digital Equipment Computer User Society) during the 1960s to 1990s. When Compaq acquired Digital in 1998, the users group was renamed CUO, the Compaq Users' Organisation. When HP acquired Compaq in 2002, CUO became HP-Interex, although there are still DECUS groups in several countries. In the United States, the organization is represented by the Encompass organization.

Notes

[1] "DEC used by Digital itself:" *PDP11 Processor Handbook* (1973): page 8, "DEC, PDP, UNIBUS are registered trademarks of Digital Equipment Corporation;" page 1-4, "Digital Equipment Corporation (DEC) designs and manufactures many of the peripheral devices offered with PDP-11's. As a designer and manufacturer of peripherals, DEC can offer extremely reliable equipment... The LA30 DECwriter, a totally DEC-designed and built teleprinter, can serve as an alternative to the Teletype."

[2] "MITRE's Project Whirlwind Computer Collection Transferred to MIT" (http://www.mitre.org/news/releases/09/whirlwind_07_01_2009.html), MITRE, 1 July 2009

[3] "Semi-Automatic Ground Environment (SAGE)" (http://www.mitre.org/about/sage.html), MITRE, 25 January 2005

[4] "TX-0 Computer" (http://www.computerhistory.org/collections/accession/XD154.75), Computer History Museum

[5] Highlights from *The Computer Museum Report* Volume 8 Spring 1984 (http://ed-thelen.org/comp-hist/TheCompMusRep/TCMR-V08.html), The Computer Museum, Boston, MA, archived at ed-thelen.org, retrieved 19 February 2010

[6] "Digital Equipment Corporation" (http://www.answers.com/topic/digital-equipment-corporation), *International Directory of Company Histories*, Volume 6, St. James Press, 1992

[7] "A Proposal to American Research and Development Corporation 27 May 1957" (http://archive.computerhistory.org/resources/text/dec/pdp-1/DEC.pdp_1.1957.102664472.pdf)

[8] Richard Best, Russell Doane and John McNamara, "Digital Modules, The Basis for Computers" (http://research.microsoft.com/en-us/um/people/gbell/computer_engineering/00000125.htm), *Computer Engineering, A DEC view of hardware systems design", Digital Press, 1978*

[9] "DEC Laboratory Module – FLIP-FLOP 201" (http://www.computerhistory.org/collections/accession/102633142), Computer History Museum

[10] *Present* 1978, pg. 3

[11] *Present* 1978, pg. 10

[12] Eastern Joint Computer Conference and Exhibition (http://www.computerhistory.org/collections/accession/102634161), official program of 1959 meeting in Boston

[13] "DIGITAL Computing Timeline, 1960" (http://vt100.net/timeline/1960.html)

[14] *Computers and Automation*, April 1961, pg. 8B

[15] "Bureau of Labor Statistics Inflation Calculator, 1961–2011" (http://www.bls.gov/data/inflation_calculator.htm)

[16] "History of Computing", Lexikon Services, ISBN 0944601782

[17] *Datamation*, Volume 5 Number 6 (November/December), pg. 24

[18] "Preliminary Specifications: Programmed Data Processor Model Three (PDP-3)" (http://www.gutenberg.org/etext/29461), DEC, October 1960

[19] Posting in "Announcements from The DEC Connection" (http://www.decconnection.org/announcements.htm), *The DEC Connection*, 14 January 2007

[20] Gordon Bell, "DIGITAL Computing Timeline, 1962, PDP-4" (http://research.microsoft.com/en-us/um/people/gbell/digital/timeline/1962-1.htm)

[21] Gordon Bell, "DIGITAL Computing Timeline, 1964, PDP-7" (http://research.microsoft.com/en-us/um/people/gbell/digital/timeline/1964-3.htm)

[22] Gordon Bell, "DIGITAL Computing Timeline, 1965, PDP-7A" (http://research.microsoft.com/en-us/um/people/gbell/digital/timeline/1964-3.htm)

[23] Eric Steven Raymond, "Origins and History of Unix, 1969–1995" (http://www.faqs.org/docs/artu/ch02s01.html), 19 September 2003

[24] Gordon Bell, "DIGITAL Computing Timeline, 1965, PDP-9" (http://research.microsoft.com/en-us/um/people/gbell/digital/timeline/1966-1.htm)

[25] DEC Advertisement, *Chemical and Engineering News*, Volume 46 (1968), pg. 85

[26] Miller 1997, pg. 452

[27] Wesley Clark, "The Linc, Perhaps the First Mini-Computer" (http://www.historyofscience.com/G2I/timeline/index.php?id=1006), From Cave Paintings to the Internet

[28] "DEC FAQ: What is a PDP-8?" (http://www.faqs.org/faqs/dec-faq/pdp8-models/section-1.html)

[29] "DEC PDP-8 minicomputer, 1965" (http://www.sciencemuseum.org.uk/objects/computing_and_data_processing/1982-960.aspx), *The Science Museum*
[30] "Internet History:1965" (http://www.computerhistory.org/internet_history/), Computer History Museum
[31] *Present* 1978, pg. 7
[32] *Present* 1978, pg. 8
[33] Miller 1997, pg. 456
[34] Miller 1997, pg. 457
[35] Gordon Bell, "DIGITAL Computing Timeline, 1964, PDP-6" (http://research.microsoft.com/~gbell/Digital/timeline/1964-1.htm)
[36] "PDP-6 Timesharing Software" (http://www.bitsavers.org/pdf/dec/pdp6/PDP-6_TimsharingBroch.pdf), DEC Publication F-61B
[37] Larry McGowan, "How the PDP-11 Was Born (according to Larry McGowan) (http://hampage.hu/pdp-11/birth.html), 19 August 1998
[38] alt.folklore.computers List of Frequently Asked Questions (http://www.netwhatever.com/faq/inicio.html#queIII15)
[39] "The day Jupiter went out of orbit" (http://books.google.ca/books?id=W7skAQAAIAAJ&q=dec+Jupiter+project+vax&dq=dec+Jupiter+project+vax&client=safari&cd=4), *Electronic Business*, Volume 10 (1984), pg. 76–79
[40] DEC Microprocessors: NVAX (1991) (http://simh.trailing-edge.com/semi/nvax.html)
[41] Olsen later claimed he was referring to home automation, see "Ken Olsen" (http://www.snopes.com/quotes/kenolsen.asp)
[42] Croxton, Greg. "DEC Robin (VT-180) & documentation" (http://www.digibarn.com/collections/systems/dec-robin/). *DigiBarn Computer Museum.* . Retrieved 21 March 2011.
[43] "The Rainbow 100 Frequently Asked Questions" (http://rainbow-100.com/faq/question/1/). *Drive W.* Approximatrix, LLC. 2009. . Retrieved 15 December 2010.
[44] Stravers, Kees. "The RX50 FAQ" (http://users.bart.nl/users/pb0aia/vax/rx50.html). *Kees's VAX page.* . Retrieved 21 March 2011.
[45] "Geek Historian". "MP01482 RX50 EngrDrws Jul82" (http://www.scribd.com/doc/881906/Tech-History-Digital-Equipment-Corporation-MP01482-RX50-EngrDrws-Jul82). *Tech History – Digital Equipment Corporation.* . Retrieved 21 March 2011.
[46] Wilson, John. "PUTR.COM V2.01" (http://www.dbit.com/pub/putr/). . Retrieved 21 March 2011. This relatively-recent work is a well-developed example of programs to enhance interchange of data between DEC formatted media and standard PC systems
[47] John Hennessy, David Patterson and David Goldberg, "Computer Architecture: A Quantitative Approach" (http://books.google.com/books?id=XX69oNsazH4C&pg=PA152), Morgan Kaufmann, 2003, pg. 152
[48] John Markoff, "Market Place; Digital Finally Follows a Trend", *The New York Times*, 16 July 1990
[49] Dileep Bhandarkar et al., "High performance issue oriented architecture" (http://ieeexplore.ieee.org/xpl/freeabs_all.jsp?arnumber=63667), *Proceedings of Compcon Spring '90*, pg. 153–160
[50] Mark Smotherman, "PRISM (Parallel Reduced Instruction Set Machine)" (http://www.cs.clemson.edu/~mark/prism.html), Clemson University School of Computing, October 2009
[51] Thomas Furlong et al., "Development of the DECstation 3100" (http://www.dtjcd.vmsresource.org.uk/pdfs/dtj_v02-02_1990.pdf), *Digital Technical Journal*, Volume 2 Number 2 (Spring 1990), pg. 84–88
[52] www.top500.org Top 10 Supercomputing Sites, November 2004 (http://www.top500.org/lists/2004/11)
[53] For in-depth articles regarding Digital technologies, refer to the archived Digital Technical Journal (http://www.hpl.hp.com/hpjournal/dtj/past.htm)
[54] Ned Batchelder and Vt100.net (http://vt100.net/dec/logo).
[55] Schein, *et al*, pp. 67, 109.
[56] Schein, *et al*, p. 233.
[57] Schein, *et al*, pp. 128, 144, 234.
[58] http://replay.waybackmachine.org/20060813131029/http://www.mentec-inc.com/
[59] "DEC, Cyrix sue Intel", by Gale Bradley and Jim Detar, *Electronic News* **43**, #2168 (19 May 1997), ISSN 1061-6624.
[60] SEC Web site (http://www.secinfo.com/dsvRx.83Pe.7.htm) retrieved 22 January 2008
[61] HP StorageWorks – Data and Network Storage Products and Solutions (http://h18006.www1.hp.com/storage/)
[62] Digital.com (http://www.digital.com/), DEC.com (http://www.Dec.com/)
[63] Digital Technical Journal – Online Issues (http://www.hpl.hp.com/hpjournal/dtj/past.htm)
[64] At least some of the research reports are available online at ftp.digital.com (ftp://ftp.digital.com/pub/DEC/), in the subdirectories WRL, SRC, NSL, CRL, PRL (see Research section). Verified July 2006
[65] dec.com (http://research.microsoft.com/Users/gbell/Digital/timeline/1985-6.htm)
[66] DECTEI-L Archives – February 1994 (#2) (http://listserv.acsu.buffalo.edu/cgi-bin/wa?A2=ind9402&L=dectei-l&T=0&P=200)

References

- (*Present*), "Digital Equipment Corporation: Nineteen Fifty-Seven to the Present" (http://research.microsoft.com/en-us/um/people/gbell/digital/DEC 1957 to Present 1978.pdf), DEC Press, 1978
- David Donald Miller, "OpenVMS Operating System Concepts" (http://books.google.ca/books?id=s5lDmf3ANhEC), Elsevier, 1997
- Alan R. Earls, *Digital Equipment Corporation*; Arcadia Publishing, 2004, ISBN 0-7385-3587-7.
- Edgar H. Schein, Peter S. DeLisi, Paul J. Kampas, and Michael M. Sonduck, *DEC Is Dead, Long Live DEC – The Lasting Legacy of Digital Equipment Corporation*; San Francisco: Barrett-Koehler, 2003, ISBN 1-57675-225-9.
- Jamie P. Pearson, *Digital At Work – Snapshots of the First 35 Years* (http://research.microsoft.com/en-us/um/people/gbell/digital/Digital at work 1992.pdf); Digital Press, 1992, ISBN 1-5555-8092-0
- Glenn Rifkin, and George Harrar, *The Ultimate Entrepreneur – The Story of Ken Olsen and Digital Equipment Corporation*; Contemporary Books, 1988, ISBN 0-8092-4559-0.
- C. Gordon Bell, J. Craig Mudge, and John E. McNamara, *Computer Engineering – A DEC View of Hardware Systems Design*; Digital Press, 1978, ISBN 0-932376-00-2.

External links

- GBell's CyberMuseum for Digital Equipment Corp (DEC) (http://research.microsoft.com/users/gbell/Digital/DECMuseum.htm)
- Rise and Fall of Digital (Equipment Corporation) (http://technikum29.de/en/devices/dec-history.shtm), a company chronicle at a German computer museum
- Ken Olsen (http://www.economicadventure.org/decision/olsen.pdf), New England Economic Adventure

DEC_T-11

The **T-11**, also known as DC310, is a microprocessor that implements the PDP-11 instruction set architecture (ISA) developed by Digital Equipment Corporation. The T-11 was code-named "Tiny". It was developed for embedded systems and was the first single-chip microprocessor developed by DEC. It was sold openly and was used by DEC in disk controllers, auxiliary processors and in the Atari System 2 arcade game system. It operated at 2.5 MHz, used a 5 V power supply and dissipated less than 1.2 W. It contains 13,000 transistors, uses NMOS logic, and was fabricated in a NMOS process.

References

- Olsen, R., Dobberpuhl, D. (1981). "A 13,000 transistor NMOS microprocessor". International Solid-State Circuits Conference Digest of Technical Papers. pp. 108–109.

MicroVAX_78032

The **MicroVAX 78032** (otherwise known as the **DC333**) is a microprocessor developed and fabricated by Digital Equipment Corporation (DEC) that implemented a subset of the VAX instruction set architecture (ISA). The 78032 was used exclusively in DEC's VAX-based systems, starting with the MicroVAX II in 1985. When clocked at a frequency of 5 MHz, the 78032's integer performance is comparable to the VAX-11/780 superminicomputer, which was introduced on 25 October 1977.[1] The microprocessor could be paired with the MicroVAX 78132 floating point accelerator for improved floating point performance. The 78032 represented a number of firsts for DEC. It was DEC's first single-chip microprocessor implementation of the VAX ISA and DEC's first self-fabricated microprocessor. The MicroVAX 78032 is also notable as it was the first semiconductor device to be registered for protection under the Semiconductor Chip Protection Act of 1984.

The MicroVAX 78032 contains 125,000 transistors on a 8.7 by 8.6 mm (74.82 mm^2) die that was fabricated in DEC's ZMOS process, a 3.0 μm NMOS logic process with two layers of aluminum interconnect. The die is packaged in a 68-pin surface-mounted leaded chip carrier.

MicroVAX 78132

The **MicroVAX 78132** (otherwise known as the **DC337**) is a floating-point coprocessor for the MicroVAX 78032 microprocessor. It interfaces to the MicroVAX 78032 via the DAL bus and a few control lines. The MicroVAX 78132 was responsible for executing 61 out of 70 floating-point instructions defined by the MicroVAX subset of the VAX ISA and accelerates nine integer instructions.

The MicroVAX 78132 has a 100 ns microcycle divided into four 25 ns clock phases which is generated by a 40 MHz input clock. It was composed of five "elements": a 67-bit fraction processor, a 13-bit exponent processor, a 1-bit sign processor, a microsequencer and a bus interface unit.

The MicroVAX 78132 contained 32,141 transistors on a die measuring 8.4 mm by 6.6 mm (55.44 mm^2). It was fabricated in DEC's ZMOS process, a 3.0 μm NMOS process with two layers of aluminum interconnect; and is packaged in a 68-pin surface-mounted leaded chip carrier.[2]

References

[1] Dobberpuhl, Daniel W.; et al. (1986). "The MicroVAX 78032 Chip, A 32-Bit Microprocessor". *Digital Technical Journal* (Digital Equipment Corporation) **1** (2).

[2] Bidermann, William R.; et al. (1986). "The MicroVAX 78132 Floating Point Chip". *Digital Technical Journal* (Digital Equipment Corporation) **1** (2).

Further reading

- Wolrich, G. et al. (October 1984). "A high performance floating point coprocessor" (http://ieeexplore.ieee.org/xpls/abs_all.jsp?arnumber=1052209). *IEEE Journal of Solid-State Circuits* **19** (5): pp. 690–696.

See also

U80701 - identical copy of the 78032 manufactured 1989/90 in East Germany.

External links

- "MicroVAX II" at The Computer History Simulation Project (http://simh.trailing-edge.com/semi/uvax.html)

StrongARM

The **StrongARM** is a family of computer microprocessors developed by Digital Equipment Corporation and manufactured in the late 1990's which implemented the ARM V4 instruction set architecture. It was later sold to Intel in 1997, who continued to manufacture it before replacing it with the XScale in the early 2000's.

DEC StrongARM SA-110 Microprocessor

History

The StrongARM was a collaborative project between DEC and Advanced RISC Machines to create a faster ARM microprocessor. The StrongARM was designed to address the upper-end of the low-power embedded market, where users needed more performance than the ARM could deliver while being able to accept more external support. Targets were devices such as newer personal digital assistants and set-top boxes.[1] [2]

Traditionally, the semiconductor division of DEC was located in Massachusetts. In order to gain access to the design talent in Silicon Valley, DEC opened a design center in Palo Alto, California. This design center was led by Dan Dobberpuhl and was the main design site for the StrongARM project. Another design site which worked on the project was in Austin, Texas that was created by some ex-DEC designers returning from Apple Computer and Motorola. The project was set up in 1995, and quickly delivered their first design, the **SA-110**.

DEC agreed to sell StrongARM to Intel as part of a lawsuit settlement in 1997.[3] Intel used the StrongARM to replace their ailing line of RISC processors, the i860 and i960.

When the semiconductor division of DEC was sold to Intel, many engineers from the Palo Alto design group moved to SiByte, a start-up company designing MIPS system-on-a-chip (SoC) products for the networking market. The Austin design group spun off to become Alchemy Semiconductor, another start-up company designing MIPS SoCs for the hand-held market. A new StrongARM core was developed by Intel and introduced in 2000 as the XScale.[4]

SA-110

The SA-110 was the first microprocessor in the StrongARM family. The first versions, operating at 100, 160, and 200 MHz, were announced on 5 February 1996.[5] When announced, samples of these versions were available, with volume production slated for mid-1996. Faster 166 and 233 MHz versions were announced on 12 September 1996.[6] Samples of these versions were available at announcement, with volume production slated for December 1996. Throughout 1996, the SA-110 was the highest performing microprocessor for portable devices.[7] Towards the end of 1996 it was a leading CPU for internet/intranet appliances and thin client systems.[8] The SA-110's first design win was the Apple MessagePad 2000.[9] It was also used in a number of products including the Acorn Computers Risc PC and Eidos Optima video editing system. The SA-110's lead designers were Daniel W. Dobberpuhl, Gregory W.

Hoeppner, Liam Madden, and Richard T. Witek.[1]

Description

The SA-110 had a simple microarchitecture. It was a scalar design that executed instructions in-order with a five-stage classic RISC pipeline. The microprocessor was partitioned into several blocks, the IBOX, EBOX, IMMU, DMMU, BIU, WB and PLL. The IBOX contained hardware that operated in the first two stages of the pipeline such as the program counter. It fetched, decoded and issued instructions. Instruction fetch occurs during the first stage, decode and issue during the second. The IBOX decodes the more complex instructions in the ARM instruction set by translating them into sequences of simpler instructions. The IBOX also handled branch instructions. The SA-110 did not have branch prediction hardware, but had mechanisms for their speedy processing.

Execution starts at stage three. The hardware that operates during this stage is contained in the EBOX, which comprises the register file, arithmetic logic unit (ALU), barrel shifter, multiplier and condition code logic. The register file had three read ports and two write ports. The ALU and barrel shifter executed instructions in a single cycle. The multiplier is not pipelined and has a latency of multiple cycles.

The IMMU and DMMU are memory management units for instructions and data, respectively. Each MMU contained a 32-entry fully associative translation lookaside buffer (TLB) that can map 4 KB, 64 KB or 1 MB pages. The write buffer (WB) has eight 16-byte entries. It enables the pipelining of stores. The bus interface unit (BIU) provided the SA-110 with an external interface.

The PLL generates the internal clock signal from an external 3.68 MHz clock signal. It was not designed by DEC, but was contracted to the Centre Suisse d'Electronique et de Microtechnique (CSEM) located in Neuchâtel, Switzerland.

The instruction cache and data cache each have a capacity of 16 KB and are 32-way set-associative and virtually addressed. The SA-110 was designed to be used with slow (and therefore low-cost) memory and therefore the high set associativity allows a higher hit rate than competing designs, and the use of virtual addresses allows memory to be simultaneously cached and uncached. The caches are responsible for most of the transistor count and they take up half the die area.

The SA-110 contained 2.5 million transistors and is 7.8 mm by 6.4 mm large (49.92 mm^2). It was fabricated by DEC in its proprietary CMOS-6 process at its Fab 6 fab in Hudson, Massachusetts. CMOS-6 was DEC's sixth-generation complementary metal–oxide–semiconductor (CMOS) process. CMOS-6 has a 0.35 µm feature size, a 0.25 µm effective channel length but for use with the SA-110, only three levels of aluminium interconnect. It used a power supply with a variable voltage of 1.2 to 2.2 volts (V) to enable designs to find a balance between power consumption and performance (higher voltages enable higher clock rates). The SA-110 was packaged in a 144-pin thin quad flat pack (TQFP).

SA-1100

The SA-1100 was a derivative of the SA-110 developed by DEC. Announced in 1997, the SA-1100 was targeted for portable applications such as PDAs and differs from the SA-110 by providing a number of features that are desirable for such applications. To accommodate these features, the data cache was reduced in size to 8 KB.

The extra features are integrated memory, PCMCIA, and color LCD controllers connected to an on-die system bus, and five serial I/O channels that are connected to a peripheral bus attached to the system bus. The memory controller supported FPM and EDO DRAM, SRAM, flash, and ROM. The PCMCIA controller supports two slots. The memory address and data bus is shared with the PCMCIA interface. Glue logic is required. The serial I/O channels implement a slave USB interface, a SDLC, two UARTs, an IrDA interface, a MCP, and a synchronous serial port.

The SA-1100 had a companion chip, the SA-1101. It was introduced by Intel on 7 October 1998.[10] The SA-1101 provided additional peripherals to complement those integrated on the SA-1100 such as a video output port, two

PS/2 ports, a USB controller and a PCMCIA controller that replaces that on the SA-1100. Design of the device started by DEC, but was only partially complete when acquired by Intel, who had to finish the design. It was fabricated at DEC's former Hudson, Massachusetts fabrication plant, which was also sold to Intel.[11]

The SA-1100 contained 2.5 million transistors and measured 8.24 mm by 9.12 mm (75.15 mm^2). It was fabricated in a 0.35 μm CMOS process with three levels of aluminium interconnect and was packaged in a 208-pin TQFP.[12]

SA-1110

The SA-1110 was a derivative of the SA-110 developed by Intel. It was announced on 31 March 1999, positioned as an alternative to the SA-1100.[13] At announcement, samples were set for June 1999 and volume later that year. Intel discontinued the SA-1110 in early 2003.[14] The SA-1110 was available in 133 or 206 MHz versions. It differed from the SA-1100 by featuring support for 66 MHz (133 MHz version only) or 103 MHz (206 MHz version only) SDRAM.[15] Its companion chip, which provided additional support for peripherals, was the SA-1111. The SA-1110 was packaged in a 256-pin micro ball grid array. It was used in mobile phones, personal data assistants (PDAs) such as the Compaq (later HP) iPAQ and HP Jornada, the Sharp SL-5x00 Linux Based Platforms and the Simputer.[16]

SA-1500

The SA-1500 was a derivative of the SA-110 developed by DEC initially targeted for set-top boxes.[17] [18] It was designed and manufactured in low volumes by DEC but was never put into production by Intel. The SA-1500 was available at 200 to 300 MHz. The SA-1500 featured an enhanced SA-110 core, an on-chip coprocessor called the *Attached Media Processor* (AMP), and an on-chip SDRAM and I/O bus controller. The SDRAM controller supported 100 MHz SDRAM, and the I/O controller implemented a 32-bit I/O bus that may run at frequencies up to 50 MHz for connecting to peripherals and the SA-1501 companion chip.

The AMP implemented a long instruction word instruction set containing instructions designed for multimedia, such as integer and floating-point multiply–accumulate and SIMD arithmetic. Each long instruction word is 64 bits wide and specifies an arithmetic operation and a branch or a load/store. Instructions operate on operands from a 64-entry 36-bit register file, and on a set of control registers. The AMP communicates with the SA-110 core via an on-chip bus and it shares the data cache with the SA-110. The AMP contained an ALU with a shifter, a branch unit, a load/store unit, a multiply–accumulate unit, and a single-precision floating-point unit. The AMP supported user-defined instructions via a 512-entry writable control store.

The SA-1501 companion chip provided additional video and audio processing capabilities and various I/O functions such as PS/2 ports, a parallel port, and interfaces for various peripherals.

The SA-1500 contains 3.3 million transistors and measures 60 mm^2. It was fabricated in a 0.28 μm CMOS process. It used a 1.5 to 2.0 V internal power supply and 3.3 V I/O, consuming less than 0.5 W at 100 MHz and 2.5 W at 300 MHz. It was packaged in a 240-pin metal quad flat package or a 256-ball plastic ball grid array.

References

[1] Montanaro, James et al. (1997). "A 160-MHz, 32-b, 0.5-W CMOS RISC Microprocessor" (http://www.hpl.hp.com/hpjournal/dtj/vol9num1/vol9num1art5.pdf). *Digital Technical Journal*, vol. 9, no. 1. pp. 49–62.

[2] "Digital targets supercharged StrongARM chip at consumer electronics market" (http://www.thefreelibrary.com/DIGITAL+TARGETS+SUPERCHARGED+StrongARM+CHIP+AT+CONSUMER+ELECTRONICS...-a017919435). PR Newswire. 1996-02-05. . Retrieved 2011-06-07.

[3] Erich Luening (1997-10-27). "Intel, Digital settle suit" (http://news.cnet.com/2100-1023-204668.html). *CNet news.com*. . Retrieved 2008-07-29.

[4] "Intel Unveils StrongARM-2 at Embedded Processor Forum"; Embedded Processor Watch #46; 5/4/1999 (http://web.archive.org/web/20071019213400/http://www.mdronline.com/publications/epw/issues/epw_46.html)

[5] Digital Equipment Corporation (5 February 1996). "Digital Targets Supercharged StrongARM Chip At Consumer Electronics Market". Press release.

[6] Digital Equipment Corporation (12 September 1996). "Digital's StrongARM Chips Pull Away in Embedded Race". Press release.

[7] Turley, Jim (27 January 1997). "Embedded Vendors Seek Differentiation". *Microprocessor Report*, pp. 16–21.

[8] "Digital's StrongARM Microprocessors Take CPU Lead in Network Client Market" (http://www.eetimes.com/electronics-news/4146617/Digital-s-StrongARM-Microprocessors-Take-CPU-Lead-in-Network-Client-Market). *EE Times*. November 18, 1996. . Retrieved March 16, 2012.

[9] Turley, Jim (18 November 1996). "Newton First Design Win for StrongARM". *Microprocessor Report*, p. 5.

[10] Intel Corporation (7 October 1998). "Intel Introduces StrongARM Products for PC Companions". Press release.

[11] http://www.mdronline.com/publications/epw/issues/epw_17.html

[12] Stephany, R. et al. (1998). "A 200MHz 32b 0.5W CMOS RISC Microprocessor". *ISSCC Digest of Technical Papers*, pp. 238–239, 443.

[13] Intel Corporation (31 March 1999). "Intel StrongARM Processor, Companion Chip Optimized For Handheld Computing Devices". Press release.

[14] Martyn Williams (14 February 2003). "Intel puts StrongArm on death row". InfoWorld.

[15] http://www.mdronline.com/publications/epw/issues/epw_42.html

[16] http://www.mdronline.com/publications/epw/issues/epw_101.html

[17] Rick Boyd-Merrit; Peter Clarke (24 July 1998). "Intel to reveal details on StrongARM chip" (http://www.eetimes.com/news/98/1018news/strongarm.html). EE Times.

[18] Prashant P. Gandhi (18 August 1998). "SA-1500: A 300 MHz RISC CPU with Attached Media Processor" (http://www.hotchips.org/archives/hc10/3_Tue/HC10.S8/HC10.8.3.pdf). Hot Chips 10.

Further reading

- "StrongARM-1500 Grapples With MPEG-2". (8 December 1997). *Microprocessor Report.*
- Halfhill, Tom R. (19 April 1999). "Intel Flexes StrongArm With New Chips". *Microprocessor Report.*
- Litch, Tim; Slaton, Jeff (March/April) 1998). "StrongARMing Portable Communications". *IEEE Micro.* pp. 48–55.
- Santhanam, S. et al. (November 1998). "A low-cost, 300-MHz, RISC CPU with attached media processor". *IEEE Journal of Solid-State Circuits*, vol. 33, no. 11. pp. 1829–1839.
- Turley, Jim (13 November 1995). "StrongArm Punches Up ARM Performance". *Microprocessor Report.*
- Turley, Jim (15 September 1997). "SA-1100 Puts PDA on a Chip". *Microprocessor Report.*
- Witek, Rich; Montanaro, James (1996). "StrongARM: A high-performance ARM processor". *Proceedings of COMPCON '96*, pp. 188–191.

Article Sources and Contributors

Daniel_W._Dobberpuhl *Source*: http://en.wikipedia.org/w/index.php?title=Daniel_W._Dobberpuhl *Contributors*: Bender235, Chuunen Baka, Dobberpuhl, Drpickem, EoGuy, Epbr123, Henriok, Niceguyedc, Philippe, SchreyP, TerraFrost, W Nowicki, 3 anonymous edits

Electrical_engineering *Source*: http://en.wikipedia.org/w/index.php?title=Electrical_engineering *Contributors*: *s IX s IX s IX*, 216.232.64.xxx, 574069, 62.163.16.xxx, =LUKE=, A bit iffy, A. Carty, AXRL, Aaron Nitro Danielson, Abdull, Acid478, AdamJacobMuller, Adambro, Adamy, Admiral Roo, Ahoerstemeier, Airconswitch, Akaustav, Alai, Alansohn, Allan McInnes, Allens, Aluvus, Andres, Andrevruas, Andrew Levine, Anger22, Antandrus, Anthonydeg, Ap, Ardric47, Artichoker, Asterion, Astral, Ateş, Avicennasis, Axyjo, Azndope, B, B Pete, B.d. cane, Baa, Babbage, Bachrach44, Barberio, Barneca, Bart133, Battleroyali, Be gottlieb, Benhoyt, Benschop, Bergsten, Berria, BillC, Biria64, Blainster, Blanchardb, Blaserules, Bobo192, Bongwarrior, Bookandcoffee, Bookofjude, Borislav, Braces anonymous, Brad101, Branddobbe, Brighterorange, Bryan, Bunchofgrapes, Burlywood, C J Cowie, CUSENZA Mario, Calabe1992, Call me Bubba, CanadianLinuxUser, Cant be prong, Careercornerstone, Caster23, Cats have this many lives, Cbdorsett, Cburnett, Cedars, Cflm001, Chaosguru, Charles Matthews, CheMechanical, Chrispreece2007, Christopher Parham, Cihan, Clear blue style, Comleck, Commander, Commander Keane, Connor J. Harrison, Conversion script, Cool Blue, Coolguy2680, Courcelles, CrazyGlu, Cremepuff222, Curps, Cyfal, CyrilB, D6, DBZROCKS, DMChatterton, DVdm, DabMachine, Dacium, Daniel Olsen, Danny, DavidLevinson, Deckiller, Dekimasu, Demmy100, Denny, DerHexer, Dgrant, Dhern029, Doc glasgow, DocWatson42, Donhalcon, Dougjih, Doulos Christos, DrKiernan, Drew335, Dspark76, Dtalexone, Dual Freq, Duk, Dylan Lake, EagleAg04, Edcolins, Editor at Large, Ee79, El C, Eleassar, Ember of Light, Emote, Epbr123, Eric-Wester, Erlyrisa, Failurerules, Fang Aili, Fingerz, Fousefields, Freakofnurture, Fredrik, Freepme, FreplySpang, Gaius Cornelius, Gamgee, Gary King, Geologyguy, Gerry Ashton, Giftlite, Gilliam, Glane23, Gracenotes, Graham87, Grey Wanderer, Grey lover, Greyerg, Greyhood, Gsonnenf, Gustavb, Guy Hatton, Guy Macon, GyroMagician, Haeleth, Haham hanuka, HappyCamper, Hectorthebat, Henrik, Herbythyme, Heron, HexaChord, Higher intellek, Hirzel, Hittykitty, Hmains, Hobartimus, Honeyonline, HowieKlein, Hughcharlesparker, Husond, Hydnjo, Hydrogen Iodide, INkubusse, Id02365, Ignatius09, Imasleepviking, Imecs, Indiedude, Iterator12n, Ixfd64, J.delanoy, JForget, Jaime-eduardo, JamesAM, Jameshiggott, JanCeuleers, Jarble, Jatkins, Jc3s5h, Jeff3000, JeremyA, Jim Douglas, Jim1138, Joanjoc, JoaoRicardo, Joelr31, JohnTechnologist, Johnothon, Johnteslade, JonRichfield, Jonathan Hall, JonathanBuck IEC, JonathonReinhart, Josh Parris, Joshschr, Jpbowen, Jpkotta, Judson, Judygreenberg, Jujutacular, JustAGal, KD5TVI, Kablammo, Katalaveno, Kbdank71, Kebeta, KerryO77, Khoikhoi, Klonimus, Knutux, Kokonuts, Kozuch, KrakatoaKatie, Kslasher, Kungfuadam, Kurykh, Kyle Barbour, Lachlancooper, Lapost, Lassa, Lchiarav, LearnMore, LeaveSleaves, Lenasarah, Leon7, Lexi Marie, Light current, Lightdarkness, Lightmouse, Ligulem, Lihmwiki, Lindosland, LittleOldMe, Logan, Loren.wilton, LouScheffer, Lupin, M.geselowitz, MSGJ, MYCOAL, Mac, Madchen Hoch, Mahanga, Mambo No. 9, Mani1, Mar4d, Marek69, Mark Arsten, MarkGallagher, MarkSutton, Markaci, Marklittledude, Markus Kuhn, Marsbound, Master of the Orichalcos, Matt Britt, Matuenih, Mausy5043, Mav, Maxamegalon2000, Mbeychok, McCart42, Mdd, Meegs, Meekywiki, Mellow Dramamine, Mentifisto, Metacomet, Mgiganteus1, Michael Hardy, Mick8882003, Mike Rosoft, Misza13, Mongol, Moreschi, Mr Mulliner, Mr Stephen, Mrg3105, Mrgates, Mrxxteamplayerxx, Mstanley103, Mudlock, Mygerardromance, Myleslong, Nakon, Namee, Naohiro19, Nashikawa, Nedim Ardoğa, Netscott, Newformat, Nibios, Nichalp, Nick Moss, Nirvelli, Nishkid64, Nixdorf, No studying for you!, No way nueve, Nobody of Consequence, Northamerica1000, Nsaa, Obli, Ojay123, Oliver202, Omegatron, Omicronpersei8, One-minus-ten, OrbitOne, OrgasGirl, Ost316, Oxymoron83, Pablothegreat85, Palakmathur, Papeschr, Patrick-br, Perl, Peter Horn, Phil Boswell, Phil Sandifer, Phild 99, Philip Trueman, Piano non troppo, Piet Delport, Pinethicket, Pjvpjv, Pluemaster, Pmonon, PoccilScript, Pokrajac, Politepunk, Pouya, Powerelectronics, Ppntori, Pravit726, Primedieval, Prolog, Pynisbird, Quadell, Quarl, Qxz, R0binh00d903, RCIWesner, RRRR, Radiojon, Randolph02, Raul654, Rawling, Rebthered, Red-E 2 Roll, Reddi, Reedy, Reindra, Remi0o, Requestion, Rhoeg, Rich Farmbrough, Rjwilmsi, Rmalloy, RobertG, Robin F., Rodney Boyd, Rosemead, Rror, Rudi the Winner, Rurparz, Ruud Koot, Ryulong, SDNick484, SMC89, Sadi Carnot, SallyForth123, Sbyrnes321, Scarian, SchfiftyThree, SchreyP, Scisaint, Scjessey, ScooterSES, Scottfisher, Searchme, Shekure, Shizhao, Shreshth91, Sietse Snel, Sjakkalle, Skier Dude, Slakr, Slipstream, Smart Viral, Snigbrook, SoEz2Hack, SoWhy, South Bay, SpaceFlight89, Spalding, Spangineer, Spinningspark, Splash, Spliced, Splintercellguy, Ssd, Ssilvers, Starblind, Stefandprins, Stegop, Stephenkirkup, Stevertigo, Stickyfox, Sunderland06, Sunray, Superspam111, Suri 100, Suruena, Svick, Swatjester, TKD, TUF-KAT, Ta bu shi da yu, Tabby, Tagishsimon, Tanglewood4, Tangotango, Taxman, Template namespace initialisation script, Thaagenson, The Photon, The Thing That Should Not Be, Thewhip, Tholme, Thu, Tide rolls, Tiles, Tim Q. Wells, Times9, Toffile, Tony1, Tpbradbury, TravisTX, Treisijs, Triwbe, Tuvas, Tv316, TysK, Ugotsrvd, Uppalapadu, Usgside, Vary, VdSV9, Veatch, Volupteer, Washburnmav, Waveguy, Wbm1058, Wdfarmer, Wiki alf, Wiki libs, WikiLeon, WikiUserPedia, Wikipedian Penguin, Wikipelli, Wikitürkçe, Wile E. Heresiarch, Wispanow, Woohookitty, Wtshymanski, XJamRastafire, XL2D, Xinit, XxTimberlakexx, Yamamoto Ichiro, Yashgaroth, Yk Yk Yk, Yvesnimmo, Zanfar, Zeeshnets, Zoicon5, Zucchini Marie, Zzuuzz, ملاع بوبحم, 999 anonymous edits

Microprocessor *Source*: http://en.wikipedia.org/w/index.php?title=Microprocessor *Contributors*: 1111mol, 16@r, 194.230.175.xxx, 209.239.196.xxx, 7, 95jb14, A little insignificant, ActiveSelective, Adamkirkman, Adityagaur 7, Aero14, Afromayun, Ageton, Ahoerstemeier, Ahunt, Alain, Alansohn, Aldie, Alexvent, Alf Boggis, Alokchakrabarti, Altes2009, Alureiter, Aluvus, Amol.gulhane07, Andrejj, Andy Dingley, Andy M. Wang, Angelic Wraith, Anil1268, Anomalocaris, Anthony.king@westerndesigncenter.com, AnthonyQBachler, Apantomimehorse, Apyule, Arch dude, Archenzo, Arisa, Arjayay, Arnero, Arun dalvi, Asnidhin, Axeman89, BBrad31, Beerden, Beetstra, Begoon, Bender235, Berek, Berles, Bigdumbdinosaur, Biker Biker, Blainster, BobKawanaka, Bobblewik, Bongwarrior, Bookandcoffee, Brainmachine, Brandon, Brouhaha, Bubba73, C.Fred, CWY2190, Calabraxthis, Caltas, Cameltrader, Can't sleep, clown will eat me, CanadianLinuxUser, CanisRufus, Canthusus, Capricorn42, Cbturner46, Cenarium, Chankey Pathak, Cherfr, Chris the speller, Circularshift, ClanCC, Clockwork Soul, Colin99, Conskeptical, Conversion script, CoolFox, Crazysunshine, Crusadeonilliteracy, Crystallina, Cxk271, DARTH SIDIOUS 2, DMahalko, DSLITEDS, DabMachine, Damueo, DanielVonEhren, Dasarianandkumar009, Davhorn, David Gerard, David Jordan, DavidCary, Db099221, Dcljr, Deagle AP, Deor, DerHexer, Deshmukhpratik, Dgw, Dispenser, Djd1219, Donreed, Dougweller, Download, DragonHawk, Ds13, Dureo, E prosser, ENIAC, EWikist, Echo95, Edd 123, Edetic, Edunsi, Elfguy, Elipongo, Elockid, EmadIV, Endothermic, Enochhwang, Enviroboy, Epbr123, Erzahler, FatalError, Felixdakat, Ferkelparade, Figureskatingfan, Fir0002, Fnagaton, Fourohfour, Frazzydee, Fyngyrz, GVP Webmaster, Gachet, Gaius Cornelius, Galoubet, Gary King, Gcalac, Geniac, Ghettoblaster, Giftlite, GlasGhost, Glen, Glenn, Gogo Dodo, Goochelaar, Gorank4, Grafikm fr, Greenrd, Grendelkhan, Grim23, Grunt, Guy Harris, Guy Macon, Gökhan, Hadal, Haham hanuka, Hanandre, Hannes Hirzel, Hans Dunkelberg, Happysailor, Hehkuviini, Henk.muller, HenkeB, Henriok, Herbertxu, Hitchhiker89, Hooperbloob, Horsten, Hpa, IMneme, IOOI, Idleguy, Igoldste, Ilya, Inkling, Insuperable, Inswoon, Iridescent, Isacp, Island, Ixfd64, J.delanoy, JJthe13, JLaTondre, Jagged 85, Jake Wartenberg, Jamo spingal, Jaranda, Jason Michael Smithson, Jauhienij, Jcanu, Jdforrester, Jeff G., Jeffreyarcand, Jeh, Jerryobject, JiFish, Jklamo, Jo7hs2, John a s, JohnCD, JonHarder, Jonabbey, Jonathan Hall, Jondel, Jossi, Jpbowen, Jpk, Jrockley, Jsc83, Juliancolton, Junexgamer, Jusdafax, Just James, Jwy, Jóna Þórunn, KLWhitehead, Karveandhan, Kateshortforbob, Kbdank71, Kerotan, KerryVeenstra, Kingpin13, Kinu, Kjkolb, Klingoncowboy4, Kmweber, KnowledgeOfSelf, Komap, Kozuch, Krótki, Kundor, Leon7, Letdorf, Letowskie, Lexein, Liao, Lights, Lijojames, Lilg132, Lindosland, Lotje, LouI, LouScheffer, Luna Santin, MFNickster, Mac, Mack2, MafiotuL, Mahjongg, Maian, Majorly, MaltaGC, Mandarax, Mani1, Maniago, Mannafredo, Mantror, Manxarbiter, Marek69, MarkSweep, MarsRover, Mato, Matsuiny2004, Matt Britt, Matticus78, Maury Markowitz, Mav, Maxis ftw, Maxx.T, Mb3k, Mckaysalisbury, Mcleodm, Mdd4696, Mehrunes Dagon, Meldor, Melongrower, Mendiwe, Michael Hardy, Mike Van Emmerik, Mikef, Milkmandan, Minesweeper, Mintleaf, Mirror Vax, Misibacsi, Mjpieters, Moggie2002, Monowiki, MooresLaw, Moxfyre, Mr random, Mramz88, Mrwrite, Muhherfuhher, Murray Langton, Musabhusaini, Mwd, Mxn, Mysid, MysticMetal, NPalmius, Nadavspi, Nandadeep11, Nansars, NapoliRoma, Nasnema, Nate Silva, Navanee2u, Navy blue84, Neckelmann, Neelix, NellieBly, NeoPhyteRep, Netsnipe, Nihonjoe, Ninly, Norm mit, NuclearWarfare, Oknazevad, OlEnglish, Oliverdl, Olivier, OohBunnies!, Orichalque, Oystein, PCHS-NJROTC, Pankaj139, Paul23234545, Persian Poet Gal, Phil websurfer@yahoo.com, Philip Trueman, Pingveno, Pixel8, Pmsyyz, Poppafuze, Prabodh1987, Premvvnc, Privatechef, Priyabhar, Proofreader77, Qbert203, QcRef87, Qkszkdl, Qwerty Binary, Qwerty189, RA0808, RJaguar3, Rada, RainbowOfLight, Ram-Man, Rama's Arrow, Randydeluxe, Raul654, Ray Van De Walker, Rayholt, Razorpit, RedWolf, Reisio, RexNL, Rich Farmbrough, RichardOSmith, Riki, Rilak, Rjhanson54, Rmadams, Roadrunner, RobWilloner, Roberrd, Robert Bond, RobertG, Robofish, Rror, Rubikcube, Rōnin, SMC1991, Salim Muhammad, Salvio giuliano, Sanchom, Sarenne, Sbierwagen, SchuminWeb, Scipius, Sclear02, Scootey, Search4Lancer, Seav, Senthilece19, SharpenedSkills, Sherool, Shirik, Shjacks45, Shriaunsh.sw, Shubhamthakare, Shvytejimas, Sietse Snel, SimonP, Snowolf, SoCalSuperEagle, Sp, Spacepotato, Spiff, Spitfire19, SplingyRanger, Squash Racket, Srce, SreekumarC, Srinivasasha, Stan Shebs, Stdazi, Stickee, Sunshine4921, Swtpc6800, Sylvain Mielot, TEG24601, Tarif Ezaz, Tasc, Thdremily1, The Anonymous One, The Thing That Should Not Be, The1physicist, TheGreenAc3, TheJosh, Thecheesykid, Theresa knott, Think outside the box, Thumperward, Thurinym, Tigga en, Tim Ivorson, Tinkar, Tobias Bergemann, Tomekire, Tony1, Top2Cat1, Toussaint, Toytoy, Tpbradbury, Tvdm, Twang, TyA, Uncle G, Uriel8, UtherSRG, ValC, Valery Beaud, VampWillow, Versus22, Vinnothkumar, Violetriga, Vishal5310, Viskonsas, Vivekyas, Vladmihaisima, Vmanthi, Vonfraginoff, Vrenator, Warpozio, Wavelength, Wayward, Wbm1058, Webshared, Wernher, West.andrew.g, WikiPuppies, Wikipelli, Windchaser, Wjw0111, Wknight94, WojPob, Woohookitty, Wtshymanski, Wutsje, Yamamoto Ichiro, Yarvin, Yes4us, Zajacik, Zundark, Zvika, 1001 anonymous edits

United_States_Department_of_Defense *Source*: http://en.wikipedia.org/w/index.php?title=United_States_Department_of_Defense *Contributors*: 134.132.115.xxx, 193.133.134.xxx, 195.92.67.xxx, 213.253.39.xxx, 62.253.64.xxx, 64.26.98.xxx, 90, A. B., AL MADI CONQUERED, AOC25, Abrio, Adam sk, Ahoerstemeier, Ajdz, Aldis90, Alerante, Ali'i, Allens, Amasurovsky, Anders.Warga, Andy Marchbanks, Anon lynx, AnthonyQBachler, Antred11, Arcane0, ArcaneO, Ashdurbat, Avial334, Awg1010, Axiome2, BD2412, BTA-Colin, Babbler, Backinbowl, Bahamut0013, Bakkouz, Basilwhite, Bazook, Bchaosf, Behun, Beland, Ben Ben, Benjamin Gatti, Benny K T Wong, Beryllium1984, Bfoaz, Bignose, BilCat, Bissinger, Blackvault, Blathnaid, BlueAg09, Bobblewik, Bobo192, Bonaparte, Bongbang, BrekekekexKoaxKoax, Brest, Briaboru, Brion VIBBER, Bsherr, Buckshot06, Buffs, Bull-Doser, Burzmali, Bvsxzeredsd, CIreland, CJKreklow, CMWrestlemaniaPunkCharlie, COGDEN, CORNELIUSSEON, CanadianLinuxUser, Canterbury Tail, CapitalR, CardinalDan, Causa sui, Ceradon, Cews, Chaser, Chemboss, Chendy, Chroode, Chrowe, Clindberg, Cokedupclown, Color-Copycat, Coma28, ComputerGeezer, Conversion script, Coredesat, Cornellrockey, Cory06612, Cubfanpgh, Cvieg, Cybercobra, DARTH SIDIOUS 2, DMCer, Dan Austin, DavidLevinson, Davidcannon, Ddye, Desertsky85451, Dhfort, DimaY2K, DinosaursLoveExistence, DocWatson42, DoubleCross, Dr badbeats, Drumsac, Dryguy, Duffman, Dycedarg, Ed Poor, Editc, Edthed, Edward, El C, ElKevbo, Eluchil404, Emedia10, Epolk, Ezhiki, Ezuvian, Facts707, Fairsing, Fakekris, Faradayplank, Fnlayson, Foreignshore, Freemason, FreplySpang, Friedo, Frmorrison, Gbrotman, GearedBull, Gilliam, Glacier109, Golbez, GoldRingChip, Graham87, Grahamdubya, Grosscha, Guanxi, Guy Harris, HJ32, Hadal, Happyme22, Harland1, Harro5, Hede2000, Hekerui, Highspeed, Hullaballoo Wolfowitz, Husnock, Hydrogen Iodide, IZAK, Ikip, IkonicDeath, Immunize, Instantnood, Ismaeel isac, Itai, J.delanoy, JDooley, JEB90, Jaan513, Jaceyblaire777, JackieDan, Jayen466, Jbou, Jbpayton, Jean15paul, Jeherbers, Jengod, Jiang, Jmtaylor90, JoSePh, John Broughton, JohnI, Johnpseudo, Jon Holly, Jon wink 12, Jonathunder, Josephf, Josher777, Jrtayloriv, KUsam, Kairologic, Kaldari, KarlFrei, Kboehrer, Keilana, Kevin Taylor, Khazar, Khoikhoi, Kier07, King Arthur6687, Kingturtle, KnightRider, Koavf, Koblentz, Kross, Kumioko (renamed), Kyle, Kyleandi, Kzollman, LP-mn, Laubach5, Laurinavicius, Lbparker40, Lebowski 666, Levineps, Lightmouse, Lihaas, Lilmzattitude305, LizardJr8, Lotje, LtNOWIS, Luigifan, Lumoy, MC10, MCB, MZM, Macdaddy, Magiddens, Magus732, Maitiu86, Maksdo, Mandarax, Marcd30319, Marek69, Marpipi, Marshcmb, MaryD99, Materialscientist, Matthew Woodcraft, Matthew238, Mbecker, Meco, Mediawoman1, MeteorMaker, Mibs, Michiel, Minesweeper, MisfitToys, Mithras6, Mlbtaz, Mmernex, Monkeybait, Moog779, Motthoangwehuong, Muj0, Mukkakukaku, Munboy, Nascar1996, Natalie Erin, Natekbowman, NauarchLysander, NawlinWiki,

Nehrams2020, Neovu79, Neutrality, New England, Nimur, Ninly, Nixdorf, Nixeagle, Noclador, Noisy, Nopetro, Nsaa, Ohconfucius, OlEnglish, Old Guard, Orlady, OverlordQ, Oyoyoyoy, PaulHanson, Paulleake, Pauly04, Pbui44, Peruvianllama, PhilKnight, Pilettes, Piotrus, Plainsong, Pmsyyz, Pol098, Postdlf, Ppntori, Professorking, Pych0Fir3, Quarl, Qwarto, R'n'B, Racepacket, Rad Racer, Ramius69, Randalllin, RattusMaximus, RicJac, RickK, Rigadoun, Rjwilmsi, Rlove, Rogerd, Ronhjones, Rougher07, Rrius, SCEhardt, Sabre51, Sardanaphalus, SarekOfVulcan, Scoutstr295, Scriberius, Sean Whitton, Searchme, Senji, Sephiroth storm, Seth Goldin, Shangrilaista, SheepNotGoats, ShiningTor, Signaleer, Skapur, Slapierre, Sliceofmiami, Spshu, Srich32977, Srikeit, Sross (Public Policy), StAnselm, Stan Shebs, Stannered, Stature791, Stephenb, Stephenw77, Streltzer, Subcontinent, SunCreator, SuperD, Superm401, Suryoye, Swarm, SwisterTwister, T geier, TJ Spyke, Tamaratrouts, Tbhotch, Tbschwartz, Template namespace initialisation script, Tetreault1, Tetromino, Tgeairn, The 888th Avatar, The Cunctator, The Thing That Should Not Be, The-G-Unit-Boss, TheAznSensation, TheCustomOfLife, Therequiembellishere, Theslider09, Thething88, Thevstyle, Thewinchester, ThinkingTwice, Thomas.freeman, ThomasOwens, TimShell, Tom harrison, Tonster, Tpbradbury, Tungka, Twsx, Uncle Dick, UrukHaiLoR, Us441, Van helsing, Vargenau, Vegaswikian, Vikramsurya, Vlad, Voidxor, WHATaintNOcountryIeverHEARDofDOtheySPEAKenglishINwhat, WSpaceport, Waggers, Wasserwig, Watcharakorn, Weedude12345, Weiners97, WhaleyTim, WhisperToMe, Wik, Wiki alf, WikiLaurent, Wishflower, WojPob, Woohookitty, Wtlegis, X!, Ximum150, YnnusOiramo, YrPolishUncle, Yuckfoo, Zap Rowsdower, Zereshk, Zoe, Zundark, Zxcvbnmmnbvcxz, 616 anonymous edits

General_Electric *Source*: http://en.wikipedia.org/w/index.php?title=General_Electric *Contributors*: -Majestic-, 1USCo, 28421u2232nfenfcenc, A Man In Black, AEMoreira042281, AMCKen, AV3000, Aboutmovies, AbsolutDan, Acdx, AdamDeanHall, AdamW, AdnanSa, Adraeus, AdventurousSquirrel, Aeons, Aerion, Afas1407, Afishler, Aflin, Aizuku, Alan Liefting, Alanlay, Alansohn, Albany NY, Alchemy 142, Alert Investor, Alex.tan, AlexLibman, AlexiusHoratius, Alison9, All Is One, Altenmann, Alvinying, Amcl, Anarchy, Andonic, Andy G, Andy Marchbanks, Angrysockhop, Anhydrobiosis, Anna Lincoln, AnnaFrance, Anniv22, Anschelsc, Antandrus, Antennaman, ApplianceRepair, April kathleen, Aquemini, Arch dude, Aross12345, Art.baron, Arthena, Arunvish, Arx Fortis, AshevillePromise, Aspects, Astellix, Aster77, Astor14, Astuishin, Athaenara, AtilimGunesBaydin, Attitude2000, Az1568, Azumanga1, B21O303V3941W42371, BRG, Barek, Bastin, Bde1982, Beetstra, Before My Ken, Beinsane, Bender235, Benzocane, BigGuns, Bigmorr, BilCat, Bingoeleven, Biosketch, Bjfg897, Blaxthos, Boarder8925, Bob98133, Boing! said Zebedee, Bongomatic, Bongwarrior, Boothy443, Born2flie, BozMo, Bp102371, Branddobbe, Brazuca100, BrekekekexKoaxKoax, BrenDJ, Brouhaha, Bschuko, Btipling, Bubba73, CDA, Caltas, Calwatch, Cannonball 248, Canthusus, Caper13, Capricorn42, CardinalDan, CaribDigita, CaseyPenk, Catapult, Ccrrccrr, Cdogsimmons, Cdonea, Ch Th Jo, Chaitanya.lala, Chendy, Cherkash, Chris the speller, Chris2311, ChrisGualtieri, Chrisieboy, Christianceo, Christopher Parham, Chrisv55, Ckatz, Classicstruggle3, CliffC, Cliffb, Codyfinke2008, Comet Tuttle, Computerhistory, Consumer friend, Conversion script, Coolhawks88, Cousteau69, Crocodile Punter, Crysb, Cyrusc, DIEXEL, DMCer, DMG413, DVD R W, DabMachine, Dachshund, DagErlingSmørgrav, Dagrak, Dalenunes, Danallen46, Dancter, DanielPenfield, Dark567, Darkpoet2, Darkwind, Dav115, Davewho2, David Delarge, David Gale, David6.jenkins@ge.com, DavidBlackwell, Davodd, Deli nk, Deltabeignet, Dieselchicken, DigbyDalton, DinosaursLoveExistence, Diomidis Spinellis, Disavian, Dizziealle, Dleonard55, DocWatson42, DocendoDiscimus, Doctor yellow, Docu, Dogru144, Download, Dpbsmith, Dragomiloff, DragonHawk, Drewk, Drmagic, Dswislow, Duffman, Dustingc, Dylan6758, E Pluribus Anthony redux, E. Ripley, Eagleamn, Earlypsychosis, Ebaksa, EdBever, EditorInTheRye, Editthesadparts, Edokter, Edward, Ehn, Eiland, Elaragirl, ElectricEye, Elipongo, Emma Frost 481, Enviroboy, Epbr123, Erikjoncas, Escape Orbit, Eustress, Evannewman830, Evans1982, Eye1ball12, F, Fairlyoddparents1234, Fearfulsymmetry, Ferfish, Filip3311, Financial Zorro, Fingers-of-Pyrex, Finlay McWalter, Firstorm, Flyguy33, Frankie12, Freecyprus, Funandtrvl, G27, GT, Gaius Cornelius, Gallantgriffin2006, Gary King, Gcm, Generalscorch, Gnowxilef, Gogo Dodo, GoingBatty, GorillaWarfare, Gr1st, Graham87, Grecko, Greenshed, Greg5030, Gregalton, Grendelkhan, Gridlock Joe, Gsarwa, Guaka, Gurch, Guy Harris, HJ Mitchell, Haham hanuka, Haliii03, Happydemon912, Harro5, Hatcat, Hatster301, Hayfordoleary, Headbomb, Heccy, Helloelan, Heywoodg, Hibernian, Hiphats, Hmmwhatsthisdo, Hoary, Hooverhigh, Horsewack, Hotwire, Htanna, Hu12, Huck57, Hunington-Horstachio, Huntster, Hydrargyrum, Ian3055, Icairns, Icanzhavegoodwiki, Icseaturtles, Idaltu, Ideal gas equation, Ief, Imgaril, Indon, Infrogmation, Inwind, Ipsherman, Ipsingh, Ireyes, Iridescent, Irishguy, Iroll, Ishi Gustaedr, J.P.Lon, J.delanoy, JGXenite, JJay, Jacksonian3623, Jamcib, JamesBWatson, Jamie jca, Jauerback, JavierMC, Jazzamac, Jeet, Jerome Charles Potts, Jerryseinfeld, Jersyko, Jibjibjib, Jkd1242, Jkeene, JoanneB, JoeSmack, Joelfurr, Jogabbeyjr, John Fuzukaka, John K, John Vandenberg, Johnanth, Johnfos, Jonathen Skews, Joseph Solis in Australia, Jprenslo, Jrennie, Jules7484, Jvcdude, K61824, Kamlanjan, Kansan, KansasCity, Karebear 1022, Karimul, Kbdank71, Kelly Martin, Keystoneridin, Killerhunk, Kingturtle, Kkm010, Klp363, Knockwood, Kozuch, Kshanks, Ktr101, Kurieeto, Kurse 2799, Kurtbw, Kyle1278, LG4761, LUUSAP, LanceBarber, Langkid1, Larrybob, Leschiffre, LethSquir, Lethe, Leuko, Levineps, Li'l Happy Ending, Ligulem, LilHelpa, Ljenkins13, Lockesdonkey, Logosys, LorenzoB, Lostintherush, Lotje, LtPowers, Lukobe, M-hwang, M.Fin.User, MK8, MONGO, MacFodder, Madhero88, Mailer diablo, Mairi, Makemi, Manzanita, Marcschulz, Mark83, Markvs88, Martin Wisse, Masque 308, Massfrequency, Materialscientist, Mathieu.petitjean, Maury Markowitz, Mayur, Mbisgaier, Mdavidson98, Meganfoxx, MeltBanana, Mentifisto, Mephistophelian, Messiisking, Mewtwowimmer, Mfourcade, Michael Hardy, Mike Dillon, Millmoss, Mimosalane, Minesweeper, Mirmo!, Mitsukai, Modster, Moo7a, Moocowsrock, Moondancer, Mr. Billion, Mr. Power-Mad!, Mr.bub, MrOllie, Mrceleb2007, Msmusiclaw, Mugunth Kumar, Mulad, Mw988, My76Strat, N328KF, N5iln, NKCTrio, Nakon, Nascar1996, Natalie Erin, NawlinWiki, Ncox, Nczempin, Necromancer, Neelix, NeilN, Nellis, Neonlite, Neutrality, Nichenbach, Nickrg3, NicoNet, Nitin nanda, NoisyJinx, Noroton, Novickas, Noybwiki, Nroberge, Nuno Tavares, Nwbeeson, Ohconfucius, Ohnoitsjamie, Ohpuu, Oihgreugihoer, Oldag07, Oleg Alexandrov, Olivier, Olly150, Onorem, Oo64eva, Orangemonster2k1, OrgasGirl, Ost316, Ottawahitech, Oxygen you can believe in., P1mpy699, PHDrillSergeant, PTSE, PadreNuestro, Pakaraki, Papriwalprateek, Patricknewyork, Patricknoddy, PaulHanson, Paulinho28, Paulzubrinich, Penale52, Petri Krohn, Phil024, Philip Trueman, Picaroon, Pig de Wig, Plasticup, Politepunk, Porqin, PostdIf, Psiphiorg, Purpleslog, Qc, Qst, QuantumOne, Quidam65, Qxz, REDyellowGreenBLUE, RS101, RTC, Rabhyanker, Radimast, RadioFan, Rberdeen, Rcawsey, RedWolf, Rediahs, Redmiders203, Rehman, Reify-tech, Rencin24, Retro00064, Revo331, Rfc1394, Rgoodermote, Rhobite, Riana, Ribonucleic, Rich Farmbrough, Richard Arthur Norton (1958-), Richard Weil, RideABicycle, Rjwilmsi, Rknasc, Roadrunner3000, Robguru, RoboAction, Robqui, Rocketman116, Rogerd, Roscoe x, Royalguard11, Rsecker, Rtcpenguin, Ruziklan, Ryan 1269, Ryan Norton, RyanGerbil10, SDC, SamJohnston, Sanjeevmittal, Sardanaphalus, Scalphunter 363, SchreiberBike, Scififreak95, Sdxvi, Sea.wolf4, Selket, Selmo, Serchdf, Serpent's Choice, Sf5xeplus, Shadowjams, Shawnc, Sheertrror, Sherool, Shizhao, SigKauffman, Sigma 7, Sitethief, Sjsharksrs, Skoosh, SkylineEvolutionIII, Slakr, Slo-mo, Smac2020, Smeaea23, Smmgeek, Solarisworld, Sophie, Soren121, Spalding, Spangineer, SpencerWilson, SriMesh, StaticGull, SteinbDJ, Stella anne victor, Stemonitis, Stephenb, Steve03Mills, Steven.dai, Stevenmitchell, Steviedpeele, Stifle, Supra guy, Sven4698e, TAnthony, TGCP, TJ Spyke, TWCarlson, Tabletop, Tasteguru, Taxhavensetc, Tcncv, Tedder, Temporaluser, Terafly999, The Flying Spaghetti Monster, The Random Editor, The wub, TheQuandry, Thehelpfulone, Therealcolletepierre, ThinkEnemies, Thomas Larsen, ThomasK, Tide rolls, Tinton5, TitaniumDreads, Tkaizan, Tkessler, Tkgd2007, ToddSweeney, Toonmore, TransporterMan, Tregoweth, Trekphiler, Trevor MacInnis, Trevorwidger, Triplight, TrisDG, Trivialist, Triwbe, Truthsniffer, Truthsort, Tumeric12, TwoOneTwo, Twsx, Tyrider07, UCFirefly, Ufteach, Umptee, Urbanrenewal, Uris, Usquam (usurped), Ustye, UtherSRG, Valentinian, Valorian, Vegaswikian, Veghead04, Veritas 0273, Viveksinha in, Vortexrealm, Vroman, Vrraybadboy4000, WAMC Pirates, WTGDMan1986, Walloon, Warren, Wassermann, Watti Renew, Wavelength, Wcquidditch, Wgungfu, Whosasking, WikiKing1234, Wikieditor06, Wikiuser100, Will Beback Auto, WillMcC, Windwardtrades, Wine Guy, Wknight94, Wongm, Woohookitty, WookieInHeat, World8115, Xavier Canals-Riera, Xienyao, Xnatedawgx, YuMaNuMa, Yurik, ZCaliber, Zachary, Zackaback, Zenosparadox, Ziiv, Zip-Zap 223, Zmiller923, Zntrip, Zr2d2, Zweldon, Zzuuzz, أحمدغامدي24., , 968 anonymous edits

Application-specific_integrated_circuit *Source*: http://en.wikipedia.org/w/index.php?title=Application-specific_integrated_circuit *Contributors*: Adrianwn, Adrory, Aleenf1, Altenmann, Arpa, Atlant, Ausinha, Avoidhours23423, Aweinstein, Banglore orifice, Barrylb, Behnam Ghiaseddin, BenTremblay, Berserkerus, Blenda, Bradams, Buster79, Bvankuik, CanisRufus, Carbuncle, Carlossuarez46, Chris hutchings, Chris the speller, Cody3019, Coleman89, Cspan64, Damian Yerrick, Danhash, Dgrant, Dicklyon, Djlipkin23, Edward, Elikcohen, Enochlau, Eras-mus, Everyking, Evil saltine, Ewlyahoocom, Fattires1993, Favonian, Femto, Floridamicro, Frostholm, Ft1, FuelWagon, Fuzzie, GB fan, GRAHAMUK, Gerry Ashton, Giftlite, Gigoptixcxsa, Glenn, Gporter1974, Gurch, HenkeB, Henriok, Hu12, ICE77, Ian Pitchford, Imran, Indon, Irondesk, Itai, J04n, Jasonsophia, Jc3s5h, Jcoffland, Jeff G., Jehochman, Jmmooney, Jon Awbrey, JonHarder, Jonathan de Boyne Pollard, Jpbowen, Kanwar rajan, Katiemurphy1, Khalid hassani, Kkmurray, Korg, Kulp, Lcapcara, Liftarn, Lorkki, LouScheffer, Lzur, Madison Alex, Maelwys, Mamidanna, Marcika, Martarius, Maxalot, Mcleodm, Mdamrau, Melvin Simon, Missy Prissy, Mogigoma, Moonman333, MrOllie, Mtmcdaid, Murugango, Nafis ru, NavarroJ, Ndokos, Nixdorf, OSP Editor, Ohnoitsjamie, Oleg Alexandrov, Overmaster net, Palica, Panscient, Parikshit Narkhede, Patrickyip, PaulN, Pdfpdf, Pdq, Pedant17, PeterBrooks, PeterJohnBishop, Phoebe, Phoenix-forgotten, Plasticup, Pradeep.esg, Pranav tailor, R'n'B, Rahulkants, Rajesh4091989, Randyest, Requestion, Rhebus, Rich Farmbrough, Richard mount, Rilak, Robofish, Rocksheavy2234, Rosenbluh, Rwender, Sanjay.iitd2004, Scientus, SelkirkRanch, Shaddack, SteveLetwin, StuGeiger, Sudhirsathiyan, Synchronism, Tagremover, Tauheed ashraf, The Anome, Thumperward, Triddle, Trjumpet, Victor Snesarev, VoxLuna, WikHead, Wikipelli, Williamv1138, WriterHound, Wsuverkropp, Xasodfuih, 281 anonymous edits

Digital_Equipment_Corporation *Source*: http://en.wikipedia.org/w/index.php?title=Digital_Equipment_Corporation *Contributors*: 62.253.64.xxx, AV3000, AdamWill, Adamantios, Aeons, After Midnight, Agentbla, Al Kossow, AlistairMcMillan, Altenmann, Amillar, Amire80, Ancheta Wis, ArnoldReinhold, Aspects, Atlant, Austinmurphy, Azikate, Badger Drink, Bbump, Bemoeial, Bender235, Bhludzin, Binarybits, Bobblewik, Bomazi, BonsaiViking, Brianski, BrokenSphere, Brouhaha, Bryanmackinnon, CWesling, CanisRufus, CarolGray, CatherineMunro, Cbh, Cbr2702, CecilWard, Cestes, Cgordonbell, Chowbok, Chris the speller, ChrisRuvolo, Christian Storm, CieloEstrellado, Cmacd123, Cminard, Cobi, Colorprobe, Computerhistory, Conversion script, Copros, Copyeditor42, Cwolfsheep, Cyanoa Crylate, DAGwyn, DMG413, DSP-user, Damian Yerrick, Daniel Pritchard, Darkguy, David spector, DavidWBrooks, Decvet, Dennis Brown, Dmsar, Dojarca, Dolovis, Don4of4, Donnymo, Dpbsmith, Dreaded Walrus, Dyl, EagleOne, Eaglizard, Edesio, Edward, Egern, Emelbee, Emperorbma, Enginear, Epbr123, Erik Corry, Evanbauer, Evil Monkey, Ewlyahoocom, Folajimi, FrankFlanagan, Funandtrvl, Gadget850, Gbuchana, Geertivp, GentlemanGhost, Gerard Czadowski, Ghettoblaster, Gihanuk, Glasnt, Greenshed, GregCutler, Guy Harris, Hairy Dude, Hansivers, Harryboyles, Hephaestos, Herostratus, Hmains, Hmich176, IMSoP, Icairns, Ideogram, IngerAlHaosului, Ingling, Ipduh, Itai, IvanLanin, JLaTondre, Jaraalbe, Jay, JayWhitney, Jedonnelley, Jerryobject, Jhobson1, Jj137, Jjzanath, Jkbw, Jll, Jnc, John Vandenberg, Jonel, Joren Six, Jp314159, Kalinrose, Kate, Ketiltrout, KnightRider, KrakatoaKatie, Lee Cremeans, Lee Daniel Crocker, Letdorf, Levin, Lightmouse, LilHelpa, Lilac Soul, Linuxlad, Litefantastic, LolaBeDen, LoopTel, Lotje, Luminifer, Lupo, M7, MITalum, ML2000, Macrakis, Maelnuneb, Martarius, Maury Markowitz, Mav, Mercurywoodrose, MichaelKirche, Mikeroetto, Mikus, Mild Bill Hiccup, Millennas, Mindmatrix, Minesweeper, Miracleworker5263, Miremare, Mirror Vax, Misternuvistor, Moink, MrDolomite, MrMichaelJKelly, Mu301, Murray Langton, Namazu-tron, NameThatWorks, Nbrouard, Neelix, NeilN, Neilc, Notheruser, NuclearWarfare, Ofiachain, Ohconfucius, One, OrgasGirl, Ortolan88, Oxymoron42, Ozga, Ozuma, Pat Berry, Paul1337, PaulLambert, Pblaauw, Peruvianllama, Petr Kopač, Pinecone, Pkjohnston, PleaseStand, Pmj, Pol098, PrintStar, QTCaptain, Quadra630, Qwertyus, Qwyrxian, RTC, RageX, Raryel, Recury, RedWolf, Reify-tech, Rhe br, Rhsatrhs, Rich Farmbrough, Rilak, Riumplus, Rjwilmsi, Robinr57, Rogerd, Ronz, Roy G, RoyBoy, Rpb01r, Rwwww, Ryulong, Sam Korn, Sburke, SchuminWeb, ShadowHntr, Sheridan, Shieber, Sillybilly, Siobhanellis, Skapur, Softy, Soulpatch, Spacepotato, Starboy444, Stassats, StaticGull, Stephen, Stevestrange, StuartBrady, Supertouch, Sven, Svick, Swirsky, Tannin, TedPostol, Tedickey, Teratornis, TerrapinDundee, Thatdog, Theinfo, Timc, Toresbe, TwoOneTwo, Twons, Unyoyega, Urbanrenewal, Urch9999, UtherSRG, Valtyr, Vanished user 5zariu3jisj0j4irj, Vkem, Vlad, Vt100, Wa3frp, Walin, Warpozio, Wavelength, Wayward, Wernher, Wgungfu, Wiarthurhu, WikiNickEN, Woohookitty, Woz2, Wtshymanski, Xnatedawgx, Youssefsan, Zippy, Zzedar, Δ, 342 anonymous edits

DEC_T-11 *Source*: http://en.wikipedia.org/w/index.php?title=DEC_T-11 *Contributors*: Rilak

MicroVAX_78032 *Source*: http://en.wikipedia.org/w/index.php?title=MicroVAX_78032 *Contributors*: DavidCary, Giraffedata, Letdorf, LouScheffer, Procolotor, Rich Farmbrough, Rilak, Wtshymanski, 4 anonymous edits

StrongARM *Source*: http://en.wikipedia.org/w/index.php?title=StrongARM *Contributors*: AxelBoldt, Chris the speller, Conversion script, Damieng, Dave clark86, David Gerard, Dicklyon, Don4of4, Dyl, Edward, Ghettoblaster, Giraffedata, Henriok, Jalwikip, Jimw338, Letdorf, Luckyz, Magus732, Marc Mongenet, Matt Crypto, Matthäus Wander, Maury Markowitz, Mirror Vax, Morkork, N328KF, Nevilley, Rada, Rilak, Solipsist, Stephen B Streater, Trevj, Unyoyega, W Nowicki, WRJ, Wernher, 36 anonymous edits

Image Sources, Licenses and Contributors

Image:Power plant.jpg *Source*: http://en.wikipedia.org/w/index.php?title=File:Power_plant.jpg *License*: unknown *Contributors*: FlickrLickr, FlickreviewR, Para, Wdwd, WikipediaMaster, Zen 38

Image:Silego clock generator.JPG *Source*: http://en.wikipedia.org/w/index.php?title=File:Silego_clock_generator.JPG *License*: unknown *Contributors*: User:AAAndrey A

Image:Faraday Cochran Pickersgill.jpg *Source*: http://en.wikipedia.org/w/index.php?title=File:Faraday_Cochran_Pickersgill.jpg *License*: unknown *Contributors*: Painted by H.W. Pickersgill, Engraved by John Cochran

Image:Thomas Edison, 1878.jpg *Source*: http://en.wikipedia.org/w/index.php?title=File:Thomas_Edison,_1878.jpg *License*: unknown *Contributors*: Brian0918, Calliopejen1, Infrogmation, Jbarta, Kelson, Makthorpe, Materialscientist, 2 anonymous edits

Image:N.Tesla.JPG *Source*: http://en.wikipedia.org/w/index.php?title=File:N.Tesla.JPG *License*: unknown *Contributors*: Blacklake, CLI, Choihei, DIREKTOR, Emerson7, GreenStork, Kilom691, Nikola Smolenski, PRODUCER, Rainman, Veliki Kategorizator, 3 anonymous edits

Image:Kilby solid circuit.jpg *Source*: http://en.wikipedia.org/w/index.php?title=File:Kilby_solid_circuit.jpg *License*: unknown *Contributors*: Texas Instruments

Image:Molnya-1 Musee du Bourget P1010442.jpg *Source*: http://en.wikipedia.org/w/index.php?title=File:Molnya-1_Musee_du_Bourget_P1010442.jpg *License*: unknown *Contributors*: User:Pline

Image:Power pole.jpg *Source*: http://en.wikipedia.org/w/index.php?title=File:Power_pole.jpg *License*: unknown *Contributors*: Courcelles, Fir0002, Hiuppo

Image:Space Shuttle Columbia launching.jpg *Source*: http://en.wikipedia.org/w/index.php?title=File:Space_Shuttle_Columbia_launching.jpg *License*: unknown *Contributors*: NASA

File:Componentes.JPG *Source*: http://en.wikipedia.org/w/index.php?title=File:Componentes.JPG *License*: unknown *Contributors*: User:Kae

Image:80486dx2-large.jpg *Source*: http://en.wikipedia.org/w/index.php?title=File:80486dx2-large.jpg *License*: unknown *Contributors*: A23cd-s, Adambro, Admrboltz, Artnnerisa, CarolSpears, Denniss, Greudin, Julia W, Kozuch, Martin Kozák, Mattbuck, Rjd0060, Rocket000, 11 anonymous edits

Image:Bayer pattern on sensor.svg *Source*: http://en.wikipedia.org/w/index.php?title=File:Bayer_pattern_on_sensor.svg *License*: unknown *Contributors*: 32bitmaschine, AnonMoos, Cburnett, W!B:

Image:Erdfunkstelle Raisting 2a.jpg *Source*: http://en.wikipedia.org/w/index.php?title=File:Erdfunkstelle_Raisting_2a.jpg *License*: unknown *Contributors*: User:Makro Freak, User:Richard Bartz

Image:F-18E cockpit m02006112600499.jpg *Source*: http://en.wikipedia.org/w/index.php?title=File:F-18E_cockpit_m02006112600499.jpg *License*: unknown *Contributors*: US Navy

Image:MEGWARE.CLIC.jpg *Source*: http://en.wikipedia.org/w/index.php?title=File:MEGWARE.CLIC.jpg *License*: unknown *Contributors*: MEGWARE Computer GmbH

File:Intel 4004.jpg *Source*: http://en.wikipedia.org/w/index.php?title=File:Intel_4004.jpg *License*: unknown *Contributors*: Original uploader was LucaDetomi at it.wikipedia

File:Z80 arch.svg *Source*: http://en.wikipedia.org/w/index.php?title=File:Z80_arch.svg *License*: unknown *Contributors*: User:Appaloosa

File:C4004.JPG.jpg *Source*: http://en.wikipedia.org/w/index.php?title=File:C4004.JPG.jpg *License*: unknown *Contributors*: Photo by John Pilge.

File:GI250 PICO1 die photo.jpg *Source*: http://en.wikipedia.org/w/index.php?title=File:GI250_PICO1_die_photo.jpg *License*: unknown *Contributors*: User:Jamo spingal

File:80486DX2 200x.png *Source*: http://en.wikipedia.org/w/index.php?title=File:80486DX2_200x.png *License*: unknown *Contributors*: User:Uberpenguin

file:United States Department of Defense Seal.svg *Source*: http://en.wikipedia.org/w/index.php?title=File:United_States_Department_of_Defense_Seal.svg *License*: unknown *Contributors*: United States Department of Defense

File:The Pentagon US Department of Defense building.jpg *Source*: http://en.wikipedia.org/w/index.php?title=File:The_Pentagon_US_Department_of_Defense_building.jpg *License*: unknown *Contributors*: "DoD photo by Master Sgt. Ken Hammond, U.S. Air Force."

File:Truman signing National Security Act Amendment of 1949.jpg *Source*: http://en.wikipedia.org/w/index.php?title=File:Truman_signing_National_Security_Act_Amendment_of_1949.jpg *License*: unknown *Contributors*: Closeapple, Infrogmation, Sreejithk2000, The gorilla, Tony Wills, TwoWings, 1 anonymous edits

File:DoD Organization March 2012.pdf *Source*: http://en.wikipedia.org/w/index.php?title=File:DoD_Organization_March_2012.pdf *License*: unknown *Contributors*: Directorate for Organizational and Management Planning/ Office of the Director of Administration and Management/Office of the Secretary of Defense

File:DoD Structure Jan2008.png *Source*: http://en.wikipedia.org/w/index.php?title=File:DoD_Structure_Jan2008.png *License*: unknown *Contributors*: Traced by T geier

File:National_Security_Agency.svg *Source*: http://en.wikipedia.org/w/index.php?title=File:National_Security_Agency.svg *License*: unknown *Contributors*: U.S. Government

File:US-DefenseIntelligenceAgency-Seal.svg *Source*: http://en.wikipedia.org/w/index.php?title=File:US-DefenseIntelligenceAgency-Seal.svg *License*: unknown *Contributors*: U.S. Government

File:Defense Logistics Agency.jpg *Source*: http://en.wikipedia.org/w/index.php?title=File:Defense_Logistics_Agency.jpg *License*: unknown *Contributors*: Original uploader was Nobunaga24 at en.wikipedia

File:DSCA.PNG *Source*: http://en.wikipedia.org/w/index.php?title=File:DSCA.PNG *License*: unknown *Contributors*: Department of Defense

File:US-NationalGeospatialIntelligenceAgency-2008Seal.svg *Source*: http://en.wikipedia.org/w/index.php?title=File:US-NationalGeospatialIntelligenceAgency-2008Seal.svg *License*: unknown *Contributors*: U.S. Government

File:DARPA Logo.jpg *Source*: http://en.wikipedia.org/w/index.php?title=File:DARPA_Logo.jpg *License*: unknown *Contributors*: DARPA

File:The Joint Staff Org Chart.jpg *Source*: http://en.wikipedia.org/w/index.php?title=File:The_Joint_Staff_Org_Chart.jpg *License*: unknown *Contributors*: Carrollkim, FlickreviewR, Lymantria, Ricjac

File:United States Department of the Army Seal.svg *Source*: http://en.wikipedia.org/w/index.php?title=File:United_States_Department_of_the_Army_Seal.svg *License*: unknown *Contributors*: U.S. Dept. of Defense

File:United States Department of the Navy Seal.svg *Source*: http://en.wikipedia.org/w/index.php?title=File:United_States_Department_of_the_Navy_Seal.svg *License*: unknown *Contributors*: US Army Institute Of Heraldry

File:Seal of the US Air Force.svg *Source*: http://en.wikipedia.org/w/index.php?title=File:Seal_of_the_US_Air_Force.svg *License*: unknown *Contributors*: US Army Institute Of Heraldry

File:Unified Combatant Commands map.png *Source*: http://en.wikipedia.org/w/index.php?title=File:Unified_Combatant_Commands_map.png *License*: unknown *Contributors*: User:Lencer

File:U.S. Defense Spending Trends.png *Source*: http://en.wikipedia.org/w/index.php?title=File:U.S._Defense_Spending_Trends.png *License*: unknown *Contributors*: Farcaster (talk) 19:44, 11 May 2010 (UTC). Original uploader was Farcaster at en.wikipedia

File:General Electric logo.svg *Source*: http://en.wikipedia.org/w/index.php?title=File:General_Electric_logo.svg *License*: unknown *Contributors*: General Electric. Original uploader was Quibik at en.wikipedia

File:Decrease2.svg *Source*: http://en.wikipedia.org/w/index.php?title=File:Decrease2.svg *License*: unknown *Contributors*: User:Sarang

File:Increase2.svg *Source*: http://en.wikipedia.org/w/index.php?title=File:Increase2.svg *License*: unknown *Contributors*: Sarang

File:GE Fairfield Entrance.jpg *Source*: http://en.wikipedia.org/w/index.php?title=File:GE_Fairfield_Entrance.jpg *License*: unknown *Contributors*: User:Ronhjones

Image:23TonGELocomotiveSRM.jpg *Source*: http://en.wikipedia.org/w/index.php?title=File:23TonGELocomotiveSRM.jpg *License*: unknown *Contributors*: User:SriMesh

Image:GeneralElectricSign.jpg *Source*: http://en.wikipedia.org/w/index.php?title=File:GeneralElectricSign.jpg *License*: unknown *Contributors*: Bubba73 (talk), Jud McCranie. Original uploader was Bubba73 at en.wikipedia

Image:LocomotiveGaugesSRM.jpg *Source*: http://en.wikipedia.org/w/index.php?title=File:LocomotiveGaugesSRM.jpg *License*: unknown *Contributors*: User:SriMesh

File:SSDTR-ASIC_technology.jpg *Source*: http://en.wikipedia.org/w/index.php?title=File:SSDTR-ASIC_technology.jpg *License*: unknown *Contributors*: User:Livewireinnovation

File:Digital 556-flattened4.svg *Source*: http://en.wikipedia.org/w/index.php?title=File:Digital_556-flattened4.svg *License*: unknown *Contributors*: User:Brianski

File:Maynard MA Clock Tower Place.jpg *Source*: http://en.wikipedia.org/w/index.php?title=File:Maynard_MA_Clock_Tower_Place.jpg *License*: unknown *Contributors*: self

File:Dec SYSTEM BUILDING BLOCKS 1103.jpg *Source*: http://en.wikipedia.org/w/index.php?title=File:Dec_SYSTEM_BUILDING_BLOCKS_1103.jpg *License*: unknown *Contributors*: Original uploader was RTC at en.wikipedia Later version(s) were uploaded by Dpbsmith at en.wikipedia.

File:Steve Russell and PDP-1 - Vintage Computer Fair 2006.jpg *Source*: http://en.wikipedia.org/w/index.php?title=File:Steve_Russell_and_PDP-1_-_Vintage_Computer_Fair_2006.jpg *License*: unknown *Contributors*: Alex Handy

File:PDP-8.jpg *Source*: http://en.wikipedia.org/w/index.php?title=File:PDP-8.jpg *License*: unknown *Contributors*: ArnoldReinhold, Bayo, EugeneZelenko, Hellisp, Methem, 1 anonymous edits

file:KA10 mod end.jpg *Source*: http://en.wikipedia.org/w/index.php?title=File:KA10_mod_end.jpg *License*: unknown *Contributors*: User:Jnc

File:AlphaServer-2100-guts.jpg *Source*: http://en.wikipedia.org/w/index.php?title=File:AlphaServer-2100-guts.jpg *License*: unknown *Contributors*: Adamantios

File:Digital Equipment Corporation 1993 logo.png *Source*: http://en.wikipedia.org/w/index.php?title=File:Digital_Equipment_Corporation_1993_logo.png *License*: unknown *Contributors*: Original uploader was Geertivp at en.wikipedia

Image:DEC StrongARM.jpg *Source*: http://en.wikipedia.org/w/index.php?title=File:DEC_StrongARM.jpg *License*: unknown *Contributors*: Idrougge, Morkork, 1 anonymous edits

MIX
Papier aus verantwortungsvollen Quellen
Paper from responsible sources
FSC® C105338

Printed by Books on Demand GmbH, Norderstedt / Germany